SCORE READING

BOOK V

Twentieth-Century Music

by

Malcolm Barry and Roger Parker
with Celia Duffy

Music Department
OXFORD UNIVERSITY PRESS
Walton Street, Oxford OX2 6DP
200 Madison Avenue, New York NY 10016

Oxford University Press, Walton Street, Oxford OX2 6DP

Oxford New York Toronto
Delhi Bombay Calcutta Madras Karachi
Petaling Jaya Singapore Hong Kong Tokyo
Nairobi Dar es Salaam Cape Town
Melbourne Auckland

and associated companies in
Beirut Berlin Ibadan Nicosia

Oxford is a trade mark of Oxford University Press

Published in the United States
by Oxford University Press, New York

© *Malcolm Barry and Roger Parker 1987*

All rights reserved. No part of this publication may be reproduced, stored in a retrieval system, or transmitted in any form or by any means, electronic, mechanical, photocopying, recording or otherwise, without the prior permission of Oxford University Press.

ISBN 0-19-321030-4

Printed in Great Britain by
Butler & Tanner Ltd, Frome and London

Preface

Our intention has been to produce representative examples of the particular difficulties presented by twentieth-century scores, and to stimulate consideration of the relationship, for the listener, between what is seen (the score) and what is heard (the music). These points will be discussed more fully in the Introduction, but we should add here a note on our criteria for selection of extracts. Our emphasis has been on major composers – which gives the book a wider appeal, and increases its possible uses – but the nature of our topic meant inevitably that the bias was almost exclusively towards highly innovative figures. While this will probably prove acceptable to most people for the period before 1945 – Mahler, Debussy, Schoenberg, Ives, Stravinsky, Bartók, Varèse, Webern, and Messiaen – the situation after that time is much more complicated. Against Cage, Boulez, Stockhausen, Penderecki, Riley, and Cardew one might well place Britten, Tippett, Shostakovich, Henze, and Dallapiccola, claiming that the second group presents a firmer, more convincing tradition for the music of the future. But our primary aim was to identify what was new from a notational point of view, and the second group, whatever its merits, has for the most part been content to leave this area unchanged. So this is not a history of twentieth-century music; nor is it a glossary of new devices and signs (though composers' tables of symbols are included where necessary).

We have also tried as far as possible to avoid the impression that the progress of twentieth-century music is some kind of steeplechase in which the hurdles become higher as time goes on. For example, we have grouped the scores chronologically by composer (except in the Appendix), and, as it happens, one of the most complex turns out to be one of the earliest (Schoenberg's Op. 16 No. 5). For similar reasons, we have not always chosen the most difficult or problematic pieces to illustrate particular genres: Webern, for example, is represented by his String Quartet Op. 28 rather than by a late orchestral work; Boulez by *Le Marteau sans maître* rather than by *Pli selon pli*.

We have provided translations of unusual and unfamiliar instruments and terms: Italian terms which are in common use are not translated, but their French and German equivalents are. Anything not included can be found in a good dictionary of music, which should be referred to as appropriate where more detail is required.

Finally we would emphasize that the commentaries are designed to be of assistance *after* the initial impression, which must be derived from hearing the extract and, indeed, the complete work.

Several people who have been of great help to us in the writing and preparation of this book deserve mention here. Philippa Whitbread of Oxford University Press initiated the project, and has been helpful and encouraging at every stage. Carol Newing, also of OUP, undertook a fair amount of rather tedious negotiation, and reacted with good humour to several last-minute changes. Although formal acknowledgement of copyright holders is made elsewhere, we must also thank the various publishers and individuals for permission to reproduce the extracts, in particular Terry Riley for supplying us with his revised performing instructions for *In C*. Jean Russell and Joan Berry showed great care and patience in typing the manuscript.

Goldsmiths' College
1982

Malcolm Barry
Roger Parker
Celia Duffy

Contents

Introduction	iv
1. G. Mahler: Symphony No. 4 in G major, First Movement (*extract*)	1
2. C. Debussy: *Prélude à l'après-midi d'un faune* (*extract*)	6
3. C. Debussy: *Jeux* (*extract*)	13
4. A. Schoenberg: Five Orchestral Pieces, Op. 16 No. 5 (*complete*)	17
5. C. Ives: Symphony No. 4, Second Movement (*extract*)	26
6. B. Bartók: Sonata for Two Pianos and Percussion, Second Movement (*extract*)	33
7. I. Stravinsky: *The Rite of Spring* (2 extracts)	37
8. I. Stravinsky: Symphony of Psalms, Second Movement (*extract*)	42
9. E. Varèse: *Ionisation* (*extract*)	46
10. A. Webern: String Quartet, Op. 28, First Movement (*complete*)	51
11. O. Messiaen: *Oiseaux exotiques* (*extract*)	57
12. J. Cage: *Music for Marcel Duchamp* (*extract*)	62
13. P. Boulez: *Le Marteau sans maître* (*extract*)	65
14. K. Stockhausen: *Kontakte* (*extract*)	69
15. K. Penderecki: *Threnody: to the Victims of Hiroshima* (*extract*)	73
16. T. Riley: *In C* (*complete*)	76
Appendix	79
17. C. Cardew: *Treatise* (*extract*)	79
18. K. Stockhausen: *Aus den sieben Tagen*, 'Unbegrenzt' (*complete*)	82

Introduction

The problems presented by the twentieth-century score are of course as great as those presented by twentieth-century music in general. The last eighty years has been a period in which, from an artistic point of view, time seems to have accelerated. 'Movements' and 'Schools' have come and gone increasingly quickly; the ensuing sense of dislocation and rootlessness among many composers becomes ever more acute. Twelve-note music, which in England was hardly mentioned above a whisper thirty years ago, is now distinctly *passé*; Hindemith, then in considerable vogue, is barely talked of today; Stockhausen, the *enfant terrible* of the 1950s and 1960s, has retreated into mysticism; Boulez has (literally) disappeared underground. In such an atmosphere, it seems quite apt that Stravinsky – the *émigré*, the Russian, the Frenchman, the American – should seem increasingly the representative composer of our age.

All this has had its effect on the way music is written down, and has presented visual challenges to the composer, performer, and even the passive listener which are as numerous as the musical styles which have evolved. But, however chaotic the situation may seem at first, there are certain general trends which may be identified as presenting particular difficulties to the musician and student used only to scores of the seventeenth, eighteenth, and nineteenth centuries.

The first of these lies in twentieth-century music's exploration of rhythm. However large and complex a nineteenth-century score may be, the reader is usually aided by the fact that it moves in regular rhythms, at a settled tempo (or with gradual changes), and with a constant time signature. The eye thus travels at a steady speed across the page. This is not the case in much twentieth-century music. Just as the harmonic certainty and stability represented by tonality were challenged at the beginning of the century, so too was the regular rhythmic practice of the previous two centuries. Rhythm became an area of experimentation, even in some cases the most important single element of a composition.

We would stress at the outset that rhythm is not in itself an element of music – it is comprised of durations (i.e. note values) grouped by metre and organized by tempo. A change in any of these elements means a change in the rhythmic language. In some cases all of these elements were altered simultaneously (e.g. in some of Elliott Carter's music); in others (there are many examples in our book) the composer concentrates on one or another.

Many of the composers represented in this book had a different approach to rhythm but, significantly, most if not all were influenced by Stravinsky. Even Bartók, whose *Allegro barbaro* of 1911 pre-dates *The Rite of Spring*, was later to acknowledge his indebtedness (especially in the Second and Third Piano Concertos), though initially it was his research into Eastern European folk music which stimulated his interest in irregular metrical groupings.

Many rhythmic devices traceable to *The Rite* have been further exploited by more recent composers: 'additive rhythms' (which are based on a small durational unit like the quaver or semiquaver, formed into irregular groupings) are, for example, a feature of Messiaen's mature style. These rhythms do not form a pulse (they pass too quickly) and so are arranged in groupings of larger units. Adding, say, a semiquaver to groupings of quavers (e.g. ♪♪♩ becomes ♪♪♪♩) produces an irregularity which can be described as 'additive rhythm'.

Simultaneously with this exploration of rhythm as a primary compositional device (as opposed to a support for harmony) came an interest in new sources of sound. The percussion writing of Bartók and the 'orchestra' of Varèse's *Ionisation* are examples of this. Inevitably the exclusive use of percussion instruments makes rhythm prominent; equally inevitably the importance of rhythm demands new sounds. These two trends in twentieth-century music are interdependent.

A further recurring element in twentieth-century music is the tendency towards diversification and fragmentation of orchestral forces. This cannot be restricted to any particular 'school' or style of composition. Although the 'standard' orchestra of the late nineteenth century was often used (sometimes much expanded) by twentieth-century composers, many of the more advanced (e.g. Mahler) tended to explore 'chamber music' groupings within that orchestra, shifting, sometimes with deliberate emphasis, from one group to another. Often the process became a structural force within the musical argument – the changes of timbre being an important key to the listener's perception of the overall musical form.

In some composers, notably those of the Second Viennese School – Schoenberg, Berg, and Webern – this process of fragmentation becomes even more extreme, with the orchestral texture reduced to a mass of individual lines, often in complex interaction. Melodic lines, traditionally given continuity and structural force by being identified with one instrument or group of instruments, were now articulated through this method of instrumental alternation, where one instrument may play as few as three or four notes before giving way to another which continues the same line. In this manner, tone-colour becomes as important an element in the melody (perhaps more important in some cases) as pitch, duration, and intensity. Schoenberg, in his *Harmonielehre* (1911), coined a term for this instrumental usage – *Klangfarbenmelodie* (literally 'sound-colour-melody'); it has become an important device for many composers and remains so even today.

This 'fragmentation' presents immediate practical difficulties for the score reader who is used only to the Classical and Romantic repertory. One can no longer rely on the backbone of a musical argument being presented on one horizontal line (almost inevitably that of the string section). The eye must learn to travel quickly from one level of the score to another, to move in and out of the texture. One element of preparation is vital: the reader must know precisely how the instruments are arranged vertically in the score, whether a piano comes below or above a harp, whether a bassoon comes above or below the horns, and so on. A careful preliminary look at the layout of each score extract may well increase the chances of its comprehension when the music is heard for the first time.

Another problem concerns the role of the performer. It is possible to regard the eighteenth and nineteenth centuries as periods in which the performer became less and less of a creative force in music. Baroque 'scores' are, in some cases, little more than basic structures upon which a piece may be improvised:

the principal example of this is the continuo part, fundamental to most Baroque music. By the time of Haydn and Mozart, even concerto soloists and singers retained their freedom only in certain areas (the cadenza being the most obvious case). In the nineteenth century, scores became increasingly detailed in their instructions to the performer, very frequently advising in detail how to *interpret* the music; and in the twentieth, many composers have seemed to tighten their grip still further, to limit ever more rigidly the interpretative role of the performer. Stravinsky has, yet again, been a leader in this respect; he advises performers to concentrate exclusively on realizing the (often extremely detailed) directions of the score – to avoid 'interpretation' at all costs: 'My music is to be "read", to be "executed", but not to be "interpreted". I see in it nothing that requires interpretation.'*

As the century progressed this situation developed still further: in Webern's mature music virtually every note has instructions on dynamics and method of attack. The early 'totally serial' music of Stockhausen, Boulez, and others takes this to the limits (some would say beyond the limits) of human capability. The earlier piano pieces of Stockhausen are good examples of this, as are the Second and Third Piano Sonatas of Boulez. The move by these composers into electronic music, though to some extent permitted only by post-World War II technological advances, was an inevitable outcome of their preoccupation with absolute control over the musical object. But this in turn provoked a reaction which dramatically reinstated the performer's creative function, as may be seen in the extracts by Cardew and Stockhausen in the Appendix.

Given the search for new resources in music – in texture, pitch-organization (harmony and melody), and instrumental and vocal sound itself – it was inevitable that new means of notation should be sought. Since 1945 in particular many composers have found conventional notation unsuitable or inadequate. Each has tended to evolve his own alternative method, often requiring pages of explanation or commentary. This has led to a bewildering variety of signs and symbols and, sometimes, entirely new approaches to the basic function of a musical score. We discuss the latter in the Appendix; in the main body of the book we concentrate on approaches which are recognizably based on conventional notation, albeit modified to take specific account of the composer's intentions.

Composers within the European tradition, e.g. Stockhausen, even when innovatory with regard to instrumentation or notation (or both), have always attempted to control, to specify, to delimit. The 'brave new world' of electronic music which Stockhausen did so much to bring about (he was joint director of the pioneering Cologne Electronic Music Studio in the 1950s) is represented by graphic symbols, giving an impression of the sound. The rest of the score is notated more conventionally. But these graphic symbols are merely the representation of a totally controlled pre-recorded type. The score thus remains as much a guide to the performers as any from the nineteenth century, but there is less specific information for the listener. The case is similar with Stockhausen's other works: his notation is always at the service of his own creative imagination. In this he may be seen essentially as the continuation of the processes described above.

Without doubt the most inventive figure in the field of notation has been John Cage. Indeed, his work has been criticized on the grounds that he is more interested in the appearance of a score than in its sound. Parallel with this visual invention has been a tireless exploration of the boundaries of music. For example, he devised the 'prepared piano', in which (as is shown in the extract we have chosen) the sound of a conventional piano is modified by inserting various objects between the strings (screws, rubber wedges, etc.). For some of his pieces Cage also devised rhythmic structures analogous to the pitch organization of Schoenberg, which imposed a pattern of metre and duration on the work. Thus, in *First Construction (in Metal)* of 1939 (a work for percussion sextet which forms an interesting comparison with Varèse's *Ionisation*), he devises a scheme based on the proportions 4:3:2:3:4. There are sixteen large sections, each of which consists of sixteen bars grouped ('phrased') in this arrangement; within these sections, instrumentation marks off each sub-section.†

This adherence to a strict scheme reflects Cage's preoccupation with Japanese philosophy, specifically Zen, in which the Will (in this case the composer's intention) is less important than the event, even if that event is partially accidental. By setting up such a rigid scheme Cage is able to by-pass invention (or, some would say, imagination): the composer's job is merely to run the system. Cage's subsequent innovations should be seen and heard in this light, especially the granting of greater freedom to the performer in his works of the 1950s, culminating in multi-media spectacles in which music is quite different from the European tradition, with its increase in precision and definition referred to earlier, because now the score ceases simply to provide information. This fundamentally challenges the notion of the score being merely subservient to the 'music'.

The post-Cage generation of American composers reflects this new attitude. Terry Riley is a good example: in common with Steve Reich, La Monte Young, and Philip Glass, he devises plans for compositions, notates them as appropriate, and (metaphorically) releases them to the world with a shrug of his compositional shoulders. Perhaps some of this informality derives from jazz, popular, and traditional music. Riley in particular has affinities with rock, while Reich has studied with a Ghanaian master-drummer, an example of the increasing influence on 'Western' music of music from other cultures. In order for his pieces to be performed, there is, however, as with Glass and Riley, reference back to conventional notation.

It is significant that the compositional developments outlined in this Introduction as having particular repercussions for the way music is written down are not primarily concerned with pitch. As we have seen, it is the tendency to bring other elements (rhythm, texture, sound resources, etc.) to the fore – in itself, one of the most important aspects of twentieth-century music – that is reflected most radically in the score. We have of course only scratched at the surface of such developments here; but it is within this framework that the various approaches to the problem of the twentieth-century score can be considered in more detail in the following extracts.

* I. Stravinsky, R. Craft, *Stravinsky in Conversation* (London, Pelican Books, 1962), p. 132.
† see P. Griffiths, *Cage* (London, OUP, 1981).

1. G. Mahler: Symphony No. 4 in G major, First Movement (*extract*).

Gustav Mahler (1860–1911) was one of the final representatives of the great Austro-German symphonic tradition, and also a profound influence on several of the most important twentieth-century composers (e.g. Schoenberg, Britten, Shostakovich). His nine symphonies (a tenth was left unfinished) and songs (many of them with orchestral accompaniment) have become increasingly popular since the 1950s and today he is one of the most frequently performed symphonic composers, both in the concert hall and on record.

The Fourth Symphony (composed 1899–1900) is often regarded as Mahler's most 'classical', both formally and in clarity of orchestral texture. The first movement is in a (highly original) sonata form, although, as may be seen, its central tonality of G major is not made clear in the first few bars. This opening also demonstrates Mahler's tendency to construct in 'blocks' — that is, to alternate short sections of material which are often characterized by their instrumentation.

It is true that the main tune is introduced by the violins (bar 3), but soon it moves to the cellos (bar 7), and then to the first horn (bars 9–10). After that we have an almost continual alternation of 'blocks', in particular when new material is introduced. Notice also the complex performing instructions, in particular with regard to dynamic level and method of attack. Mahler's willingness to deploy instruments for very short passages, sometimes merely to strengthen or colour a tiny motif (the bassoons 8 bars after Fig. 1, the clarinet 2 bars later), is also noticeable. Perhaps the route to Schoenberg's *Klangfarbenmelodie* (see Introduction) was already mapped out, even at this stage of Mahler's career.

The extract comprises the major part of the exposition. We can identify an obvious 'second subject' in the cello theme beginning at Fig. 3, in lyrical contrast to the opening violin theme; but the intervening passages give an impression less of transition than of further, sharply differentiated, thematic statements, particularly in the passage beginning at Fig. 2. This tendency to set apart the various sections is quite at one with the 'block construction' noted above, and adds much to the overall character of the movement.

INSTRUMENTS

Pauke kettledrum
Schelle bell(s)

TERMS

Bedächtig. Nicht eilen With deliberation. Do not hurry
Vorschläge sehr kurz acciaccaturas very short
Recht gemächlich quite leisurely
Haupttempo the main tempo
Etwas zurückhaltend holding back a little
zu 2 both (players)
NB. **Die Bezeichnung / zwischen 2 Noten bedeutet: glissando** the sign / between two notes indicates *glissando*
nicht geth. not *divisi*
geth. *divisi*
Dpplgr. double stopping
Hauptzeitmass the main tempo
in B (clarinets) change to B♭
Frisch lively
kräftig vigorously
3. Ob. nimmt Engl. Horn 3rd oboe takes cor anglais
Breit gesungen broadly singing
Ton! particularly sonorous
G. Saite (play on the) G string
schwungvoll energetically
Plötzlich langsam und bedächtig suddenly slow and with deliberation
Etwas eilend hurrying a little
Wieder gemächlich leisurely again
Fliessend flowing
keck cheeky

SYMPHONIE № 4.

I.

Gustav Mahler.

2. C. Debussy: *Prélude à l'après-midi d'un faune* (*extract*).

Claude Debussy (1862–1918) is regarded by many as chief claimant to the title 'father of modern music'. In his explorations of 'non-functional' harmony (where the sound-colour of chords becomes more important than their tonal function), of new formal designs and instrumental and orchestral sonorities, he has been a basic influence on many of the leading twentieth-century composers, Stravinsky, Messiaen, and Bartók being only the most prominent. Although he produced one full-length opera, *Pelléas et Mélisande* (1902), the majority of his works are either orchestral or for solo piano.

The *Prélude à l'après-midi d'un faune* (Prelude to the Afternoon of a Faun), composed between 1892 and 1894, is the first of Debussy's major works and established his reputation as one of the most significant originators of 'modern music'. As Pierre Boulez has remarked: 'the flute of the *Faune* brought new breath to the art of music. . .and just as modern poetry surely took root in certain of Baudelaire's poems, so one is justified in saying that modern music was awakened by *L'Après-midi d'un faune*'.* This is partly because of the piece's shifting tonality (the vague sense of key produced by the tritone-bounded melody of the opening, for example, or the unresolved dissonances which occur later); and partly because of the overall effect of 'improvisation', the comparative lack of formal definition or obvious thematic development. There is also a fluidity of rhythm (also found in the opening bars, with the ties, the alternation of semiquavers and semiquaver triplets, and the continual tendency of melodies to come in on the off-beat: see, for example, the horn melody at bar 4, the oboe at bar 14) and a concern for orchestral sound as an end in itself.

This extract comprises the first main section of the piece, together with part of the second section. Notice how each return of the initial flute melody (at Figs. 1 and 2, and 6 bars after Fig. 2) has a slightly different harmonization and orchestral texture – not enough to suggest 'development' in the nineteenth-century sense, but just enough to add to the initial tonal ambiguity. The first firm cadence (into B major) does not occur until the end of this section (1 bar before Fig. 3). It is typical of the importance Debussy placed on orchestral sound that the second section is differentiated from the first primarily by its orchestral texture: the strings take off their mutes, pizzicatos appear, and the new, more transparent texture is emphasized by the use of a solo clarinet instead of the flute of section 1, although the harp is prominent in both sections. The initial concentration on the lower register of the flute itself shows Debussy's ear for precise sonorities.

* P. Boulez, 'Modern Music Begins', from *Notes of an Apprenticeship*, trans. H. Weinstock (New York, Knopf, 1968), pp. 344–5.

INSTRUMENTS

Cors à pistons en Fa french (valve) horns in F
Cymbales antiques 'antique' cymbals
Harpes harps
1ʳᵉ accordez La♯-Si♭ (etc.) first harp tune A sharp to B flat, C sharp to D flat, E natural to F flat, G sharp to A flat
Alto viola

TERMS

Très modéré very moderately
doux et expressif softly and expressively
sourdine mute
sur la touche on the fingerboard
position nat. normal position
légèrement et expressif lightly and expressively
ôtez vite les sourdines quickly take off mutes

PRÉLUDE
à L'Après-Midi d'un Faune

CLAUDE DEBUSSY
1862-1918

3. C. Debussy: *Jeux* (extract).

Jeux, written in 1912, was Debussy's last major orchestral work. It was composed as a ballet score for the famous Russian impresario, Diaghilev (who was also associated with Stravinsky's early ballets — see No. 7). The piece's most important feature is a negative one — its lack of a traditional formal or thematic structure — but the attention to timbre as a structural force in some way compensates for the absence of conventional 'signposts'. The ballet was staged on a tennis court, with three players: as one commentator has remarked, although the amorphous musical definition of the work does not 'help to get the ball over the net, [it is] what makes the game worth watching'.

The extract is in three sections differentiated by tempo: two slow sections, characterized by long-drawn-out string pedal-notes and woodwind chords, frame a *Scherzando*, in which tiny motifs are thrown between various elements of the orchestra, with prominent use of the percussion. The delicacy and originality of Debussy's orchestral thought can be seen in the opening bars, which combine a solo horn with harp harmonics, or in the *Scherzando*, where the 'mosaic' effect (immediately obvious to the eye) is maintained until the final bars.

The speed and extreme fragmentation of the *Scherzando* section may present the score-reader with some initial difficulties. It is important to locate the precise position of certain 'key' instruments, particularly in the percussion and wind sections (the tambourine, bassoon, and horns for example). But the sparseness of Debussy's orchestral texture (he remarked of this piece that his aim was to create an orchestra 'without feet') is certainly an advantage, as one can see clearly the progress of the motifs as they travel through the texture.

INSTRUMENTS

Petites flûtes piccolos
T. de B., *tambour de Basque*, tambourine
Timb. timpani

TERMS

Très lent very slowly
1º sons d'écho first (horn) with echoing sounds
2ᵉ moitié second half (of 2nd violins)
4 pup. Soli 4 desks solo
TOUS all
doux et rêveur softly and dreamily
bag. de Timb. with kettledrum sticks
sur le chevalet on the bridge
léger, marqué lightly, *marcato*
un peu en dehors brought out a little
Mouv^t. du Prélude at the speed of the Prélude
sur la touche on the fingerboard
sourdines mutes
ôtez les sourdines take off mutes

Reprinted by permission of Durand et Cie, Paris

4. A. Schoenberg: Five Orchestral Pieces, Op. 16 No. 5 ('Das obligate Rezitativ').

Arnold Schoenberg (1876–1951) was one of the great musical innovators of the twentieth century. His earliest compositions expanded and extended the Austro-German musical tradition of Brahms, Wagner, and Mahler, in particular exploring ever more extreme areas of chromaticism. The Op. 16 set of Orchestral Pieces (1909) are among the first 'atonal' works (i.e. works is which key centres are no longer an important structural force). The atonal music of this period, whether by Schoenberg or his pupils Berg and Webern, tended to be based on a text, or to be short in duration, or both. The absence of the organizing power of tonality was keenly felt by these composers and resulted in a paucity of extended orchestral or instrumental compositions.

Perhaps as a reaction to this, in the early 1920s, Schoenberg devised his 'method of composing with twelve notes', sometimes known as 'serialism' or 'dodecaphony', in which all twelve notes of the chromatic scale have a fixed and equal place in the compositional framework.

No. 5 of Schoenberg's Op. 16 is perhaps the most complex and radical of the set. Some of the other pieces make use of conventional musical devices – ostinato, pedal points, formal repetition and development – which help to 'anchor' the music around pitches, if not to give it a firm key sense. But in No. 5 there seems no identifiable form, just a melodic line which continuously unfolds against a highly contrapuntal background. We might tentatively relate this to the title, literally 'The Obbligato Recitative', which implies a stream of rhythmically and formally free music, although Schoenberg himself was not fully convinced that his title was a good idea.

'Das obligate Rezitativ' is undoubtedly one of the hardest pieces to follow in this collection. Though the *Klangfarbenmelodie* (see the Introduction) is not as consistent as in Op. 16 No. 3, entitled 'Farben' (Colours), the main melodic ideas move continually from one instrument to another. The problem is compounded by the piece's considerable contrapuntal activity. It is not at all unusual for three or four separate events to sound simultaneously (follow, for example, the various strands in the opening bars), and there is, at least as far as standard nineteenth-century practice goes, relatively little doubling of instruments or instrumental groups. The difficulty of identifying 'lines' within this music was recognized by Schoenberg, and to assist readers and performers, he invented the symbol ⊢ ̄ to signify that a line contained within the brackets is a 'principal part' (the German term is *Hauptstimme*). Though not seen in this extract, a further symbol ⋁ ̄, indicating a 'secondary part' (*Nebenstimme*) is also used by Schoenberg and his disciples. A glance through the piece, following the progress of the *Hauptstimme* through this complex texture, is perhaps the best preparation for listening with the score.

Attentive listening will suggest a structure based on factors other than pitch, for example the two dynamic climaxes, which form two landmarks for the listener, and Schoenberg's characteristic use of extreme and sudden dynamic contrasts, as at Fig. 6, where there is an abrupt change from *ff* to *pppp* in the middle string parts. The piece thus marks a transition from music based on pitch to music based on texture, the latter a concept which was to become much more important after 1945, as will be seen in later extracts.

INSTRUMENTS

Kleine Flöte piccolo
Posaunen trombones

TERMS

Bewegte Achtel agitated quavers
⊢ ̄ ⊤ (*Hauptstimme*) main voice
mit Dämpfer with mutes
ohne Dämpfer without mutes
Mittelstimme middle voice
zart tender
Baßstimme bass voice
dünn thin
voll full
sehr zart und hell very tender and bright
G Saite G string
gedämpft muted
Echotonartig like an echo
zusammen together
Nebenstimme secondary voice
leicht, dünn light and thin
II. m. D. 2nd (player) with mute
so schwach wie möglich as delicately as possible
am Steg on the bridge
sehr gebunden very smooth
weich mellow
Dämpfer weg take off mutes
sehr ruhig steigern; ohne Beschleunigung get louder very gradually without speeding up
Flatterzunge flutter-tongued
trem. am Steg tremolo on the bridge
nicht gebunden not smooth
zart tender
***p* aber breit im Ton** *piano* but broad in sound
III. gr. Fl. nimmt II. Piccolo 3rd flute takes 2nd piccolo
deutlich (Solo!) distinct (Solo!)
Schalltrichter hoch hold up the bell
3 fach get. divide into three
Dämpfer aufsetzen put on mutes

V
Das obligate Rezitativ

© Copyright 1922 by C. F. Peters, London, Frankfurt & New York
Reprinted by permission of the Publishers

5. C. Ives: Symphony No. 4, Second Movement (*extract*).

The American composer Charles Ives (1874–1954) is one of the most extraordinary and original of the twentieth century. His father, George E. Ives, was a bandmaster who encouraged his children in a broad range of music activities, often 'stretching their ears' by unconventional means: Charles Ives recalled later how his father would have them sing a tune in one key while he accompanied in another. Though Ives had some professional training he never forgot this early experience, and remained throughout his composing career an unselfconscious but thoroughgoing innovator.

The Fourth Symphony (completed in 1916) is one of Ives's most important and representative pieces. In many senses it is a culmination of his work (he stopped composing in the early 1920s) and, as if to emphasize this, he quotes from no fewer than sixteen of his previous compositions, not to mention *Camptown Races* and *Yankee Doodle*. As with much of Ives's music, it also has an extra-musical element, a general 'programme' which the composer later described as 'the searching questions of What? and Why? which the spirit of man asks of life. This is particularly the sense of the *Prelude*.' Subsequent movements provide 'answers' to these questions. In the second movement, 'an exciting, easy and worldly progress through life is contrasted with the trials of the Pilgrims in their journey through the swamps and the rough country. The occasional slow episodes – Pilgrims' hymns – are constantly crowded out and overwhelmed by the former.'

This programmatic content determines the musical form and provides a useful guide to listening: the movement proceeds by abrupt contrasts of material (providing yet another example of 'block-form'), with the opening material taking over towards the end of the extract, where the cellos and basses with their downward glissandos are joined by a heavily syncopated piano ostinato from Fig. 4, both these elements gradually wearing down the violin melody – the Pilgrims' hymn. The clearest example of the abrupt juxtaposition of material mentioned above occurs at the beginning of the extract: the first page of the score, with its noisy, chaotic music, is followed by the delicate sonority of upper strings (playing harmonics and quartertones, marked *ppp* and *pp*), 'scarcely audible' piano, and a flute melody (see Fig. 2 onwards). Examples of Ives's careful attention to sonority and sometimes unconventional directions abound in this extract (see, for example, the use of a very large percussion section, the extreme dynamics (triangle *pppp* in bar 2), and the various types of string writing (harmonics, glissandos, etc.)).

As a glance at the first page of the score will reveal, the most formidable problem for performers (and score-readers) is the rhythm. Two conductors are required, and the meshing of speed and time signatures (see the directions to conductors on the first and final pages of the extract) is compounded initially by the extreme complexity of much of the rhythm and further by such directions as 'accel. freely' (see piano and bassoon on the first page), giving the individual instrumentalist a certain amount of freedom. Henry and Sidney Cowell, the composer's early biographers, tell the following story about this movement:

At the sight of the second movement of the Fourth Symphony, every orchestral conductor exclaims at once: 'Impossible to conduct the piece!' When in early 1927 the score was shown to Eugene Goossens, he said exactly the same thing, but thereafter he proceeded differently from other people, for he wound a towel about his head, drank gallons of coffee, sat up nights, learned the score, and found a way to conduct it successfully in public.*

The score-reader's best preparation is to locate the instruments that play loudly on the first page: the bassoon and double basses (at opposite ends of the score) and then the very loud brass and piano section (roughly in the middle at Fig. 1). With the *Largo* indication at Fig. 2 the music becomes much easier to follow.

For a composer as isolated as Ives seems to have been, much of this music is remarkable in its anticipation of future developments in twentieth-century music, in particular the American experimental attitude and composers such as Cage.

---

* H. and S. Cowell, *Charles Ives and his Music* (London, OUP, 1955), p. 165.

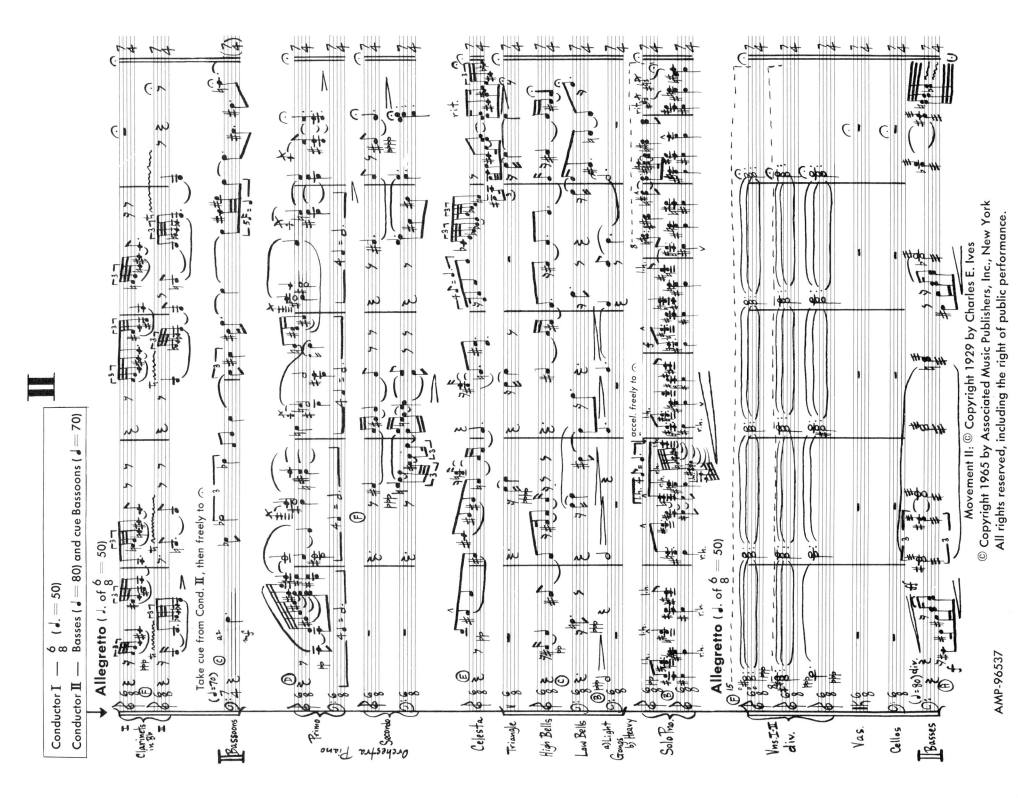

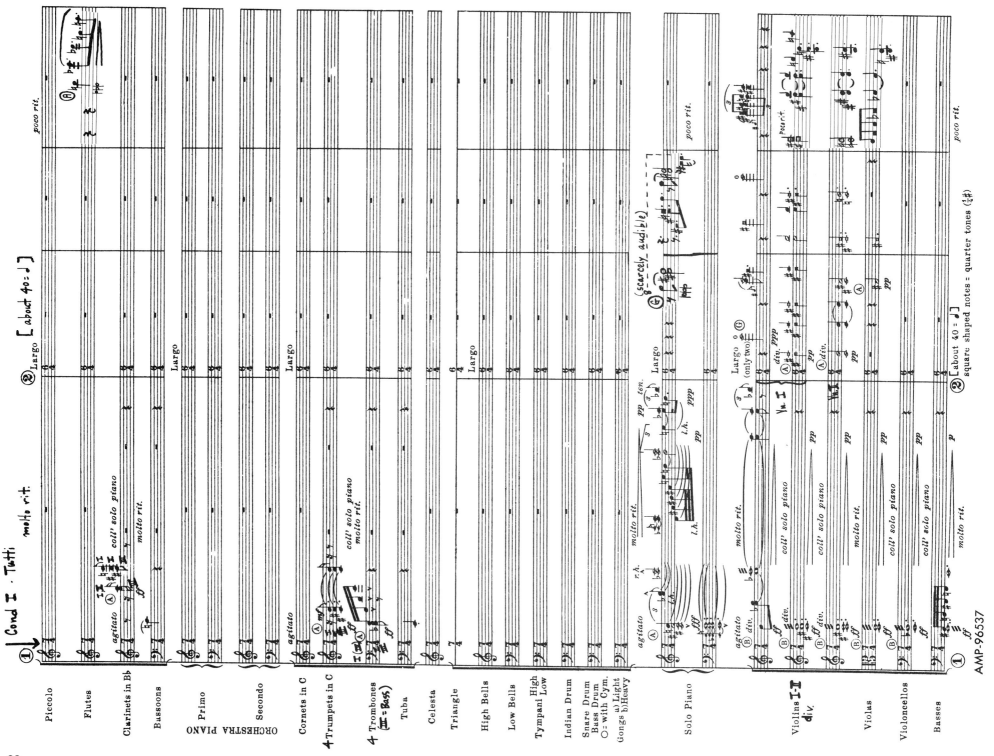

molto rit.
a tempo
Piccolo
Flutes
Clarinets in B♭
Bassoons
ORCHESTRA PIANO
Primo
Secondo
Cornets in C
Trumpets in C
Trombones
Tuba
Celesta
Triangle
molto agitato
High Bells
(scarcely audible)
Low Bells
Tympani High Low
Indian Drum
Snare Drum
Bass Drum
O = with Cym.
B. Dr.
Gongs a) Light b) Heavy
molto agitato
l. h.
molto rit.
Solo Piano
a tempo
(a♯ 8va basso only)
Violins I (2 only)
Violins I (The others)
Violins II
Violas
Violoncellos
Basses
molto rit.

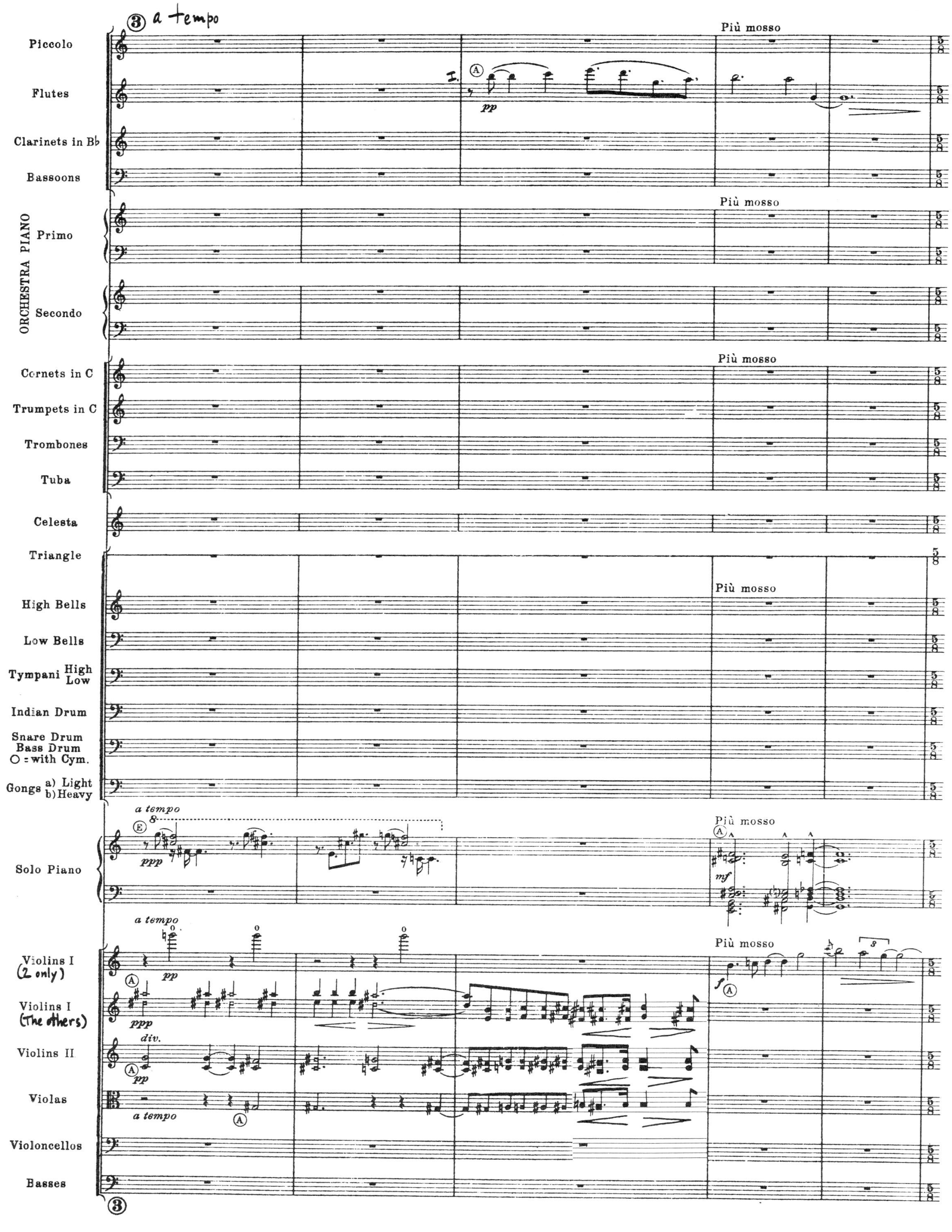

[starting about ♩= 50 and gradually faster and louder]
Piccolo
Flutes
Clarinets in B♭
Bassoons
ORCHESTRA PIANO
Primo
Secondo
Cornets in C
Trumpets in C
Trombones
Tuba
Celesta
Triangle
High Bells
Low Bells
Tympani High Low
Indian Drum
Snare Drum
Bass Drum
O = with Cym.
Gongs a) Light b) Heavy
Solo Piano
Violins I
Violins II
Violas
Violoncellos
Basses
pp
ppp
(only 2 or 3 Violins continue here)
div.
gliss.
(non gliss.)
p
[starting about 50 = ♩ and gradually faster and louder

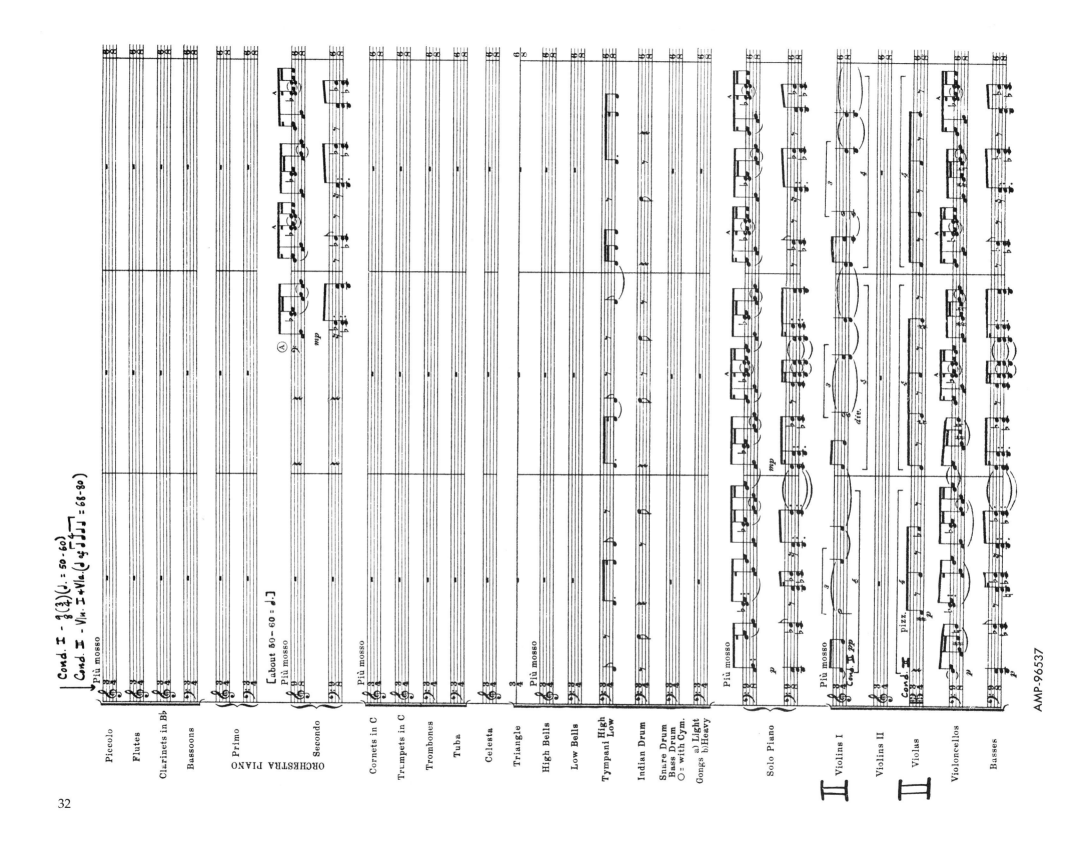

6. B. Bartók: Sonata for Two Pianos and Percussion, Second Movement (*extract*).

Béla Bartók (1881–1945) was one of the most important figures in twentieth-century music. His considerable work in the field of ethnomusicology, in particular his research into Hungarian and Romanian folk music, was an important influence in the formation of his mature style, especially with regard to rhythm and melodic forms. But Bartók was by no means isolated from the central traditions of European art music, and he was considerably influenced by Debussy and Richard Strauss. He wrote an opera, *Bluebeard's Castle* (1911), several ballet scores, various orchestral and choral works, and a substantial body of chamber music. His six string quartets lie at the centre of the modern chamber repertory.

Bartók's Sonata for Two Pianos and Percussion (1937) is a fine example of his mature style, with driving rhythms and intricate contrapuntal sections. It shows a firm, highly original attitude to overall tonal structure in the two outer movements, and a characteristic depiction of nocturnal sounds in the central *Lento*. The mixture of percussion and pianos brings into relief the fact that Bartók was primarily attracted to the piano for its percussive quality (as were many twentieth-century composers, e.g. Ives as shown in the previous extract). From the opening bars of the first movement, where a low timpani roll merges subtly into the piano theme, there is constant creative interchange between these two elements.

The opening of the second movement is pure rhythmic notation. It is worth reading it through and then trying to imagine how this first line will sound. If nothing else, this will prepare one for the notation of the next extract, Varèse's *Ionisation*. The overall form of the movement is ternary: here we have the whole of the first A section and a portion of B. In the former, the complex percussion pattern is overlaid by the pianos at their most sustained (note the precise instructions for the percussion, e.g. for the hitting positions of side drum and cymbal); in the latter (bar 28 onwards) – the 'nocturnal' evocation – the percussive potential of the keyboard instruments is brought into full prominence, pianistic imitations of the gong and xylophone eventually being complemented by the actual sounds of these instruments (although the tam-tam (a large gong) is present from bar 37). At the climax (bar 45) the xylophone takes over the semiquaver material which is soon transferred to the timpani (bar 48). From this point there is extreme contrast of registers between the xylophone, now following the new chromatic melody of piano I, and the timpani, which retain the semiquaver figure. The section is a clear example of the interchange of pianos and percussion in terms of both thematic material and instrumental timbre.

TERMS

Side Drum c.c. (con corde) with snares
Side Drum s.c. (senza corde) without snares

©Copyright 1942 by Hawkes & Son (London) Ltd.
Reprinted by permission of Boosey & Hawkes Music Publishers Ltd.

7. I. Stravinsky: *The Rite of Spring* (*2 extracts*).

Igor Stravinsky (1882–1971) is one of the major figures (many would say *the* major figure) of twentieth-century music. His career may be divided into three periods: in the first (up to *c*. 1918) his music shows the powerful influence of his early upbringing and music education in Russia; in the second (*c*. 1918–1951), a so-called 'neo-classical' period, we find a return to the classical ideals of balance and symmetry, as well as to certain forms and gestures associated with music of the eighteenth and nineteenth centuries; in the third he gradually began to evolve an extremely individual approach to Schoenberg's twelve-note method. But these changes of style can be overstressed: Stravinsky was plainly too original and too fine a composer to allow whichever form or genre he took up to obscure his essential musical personality.

The Rite of Spring (1913) is one of the most famous pieces of twentieth-century music. It provoked a riot on its first performance in Paris (though this was probably due more to Nijinsky's choreography than to the music) but has since become a standard orchestral show-piece. Although *The Rite* is highly innovative in practically every direction – harmony, melody, orchestral technique, etc. – it is the work's approach to rhythm which has perhaps excited most comment. Because of this piece's considerable importance, we have chosen two extracts – one from the beginning and one from the end.

The first extract immediately reveals an original orchestral conception – a texture made up of a trio of wind instruments and a horn, with a solo bassoon placed at the very top of its range – an unusual sonority. The extract abounds in unusual orchestral effects but we might also mention the use of six solo double basses later on at Fig. 10. As one can see, the time signature changes often; we inevitably gain the impression that the barlines are put in primarily for readability and carry little of their conventional stress function. The rhythmic irregularity and ambiguity is heightened by Stravinsky's melodic writing, with tunes which move obsessively around a small number of notes, often stating melodic progressions in more than one rhythmic context.

Another noticeable feature is Stravinsky's frequent use of superimposed layers of different metres – at Fig. 7, for example, the alto flute, first oboe, clarinets, and solo cello are playing against the indicated time signature: only the second oboe and, 3 bars later, the cor anglais uphold the ⅔ pulse. There are many other examples of this 'polymetrical texture' in the work: in this instance it derives from the free counterpoint (or 'free association') of melodic ideas derived from folk-song types.

Stravinsky's ear for precise sonorities – a characteristic shared with Debussy (see extracts 2 and 3 above) – is also evinced in the first extract, e.g. the alto flute at Fig. 6, the division of the doublebasses into six parts at Fig. 10. Again this contributes to a textural effect rather than one based on pitch (cf. Schoenberg, extract 4).

The second extract is a clear example of Stravinsky's metrical practice. Taken from the climactic 'Sacrificial Dance of the Chosen Virgin', it shows the changing of time signatures in almost every bar, the use of the whole orchestra as a percussion section (no beautiful tone on the violins here), and Stravinsky's tendency to score in blocks.

This is by no means an easy score to follow and we would stress again the importance of careful preparation: a secure knowledge of the relative positioning of the instruments within this large orchestra is essential. As a first step the reader should follow the bass line (in timpani and double basses) and see how the various sections (e.g. Fig. 199 to the bar before Fig. 200) are constructed using it as a foundation.

INSTRUMENTS

C. ing. cor anglais
Cl. picc. in Re sopranino clarinet in D

TERMS

un peu en dehors brought out a little
sul ponticello on the bridge
Descendez le 'la' un demi-ton plus bas
  tune the A (string) a semitone lower

The Rite of Spring
Le Sacre du Printemps

First Part
ADORATION OF THE EARTH

Première Partie
L'ADORATION DE LA TERRE

IGOR STRAVINSKY
Revised 1947
New edition 1967

INTRODUCTION

© Copyright 1921 by Russischer Musikverlag
Copyright assigned 1947 to Boosey & Hawkes, Inc. for all countries
Reprinted by permission of Boosey & Hawkes Music Publishers Ltd., London

8. I. Stravinsky: Symphony of Psalms, Second Movement (*extract*).

The *Symphony of Psalms* (1930) comes during what is usually termed Stravinsky's 'neo-classical' period (see introduction to extract No. 7). The 'classicism' of this movement may immediately be seen in its fugal form and the restraint and clarity of its instrumental textures; but similarities with *The Rite* are perhaps just as obvious. Again we have a highly original orchestral conception (the restriction to treble instruments of the opening section and the absence of violins and violas throughout); melodic lines which tend to move in a limited range and to repeat progressions; and (though clearly in less extreme form) frequent disruptions of regular rhythm. Stravinsky's omission of the 'expressive' upper strings contributes further to a highly personal and very restrained musical language.

Stravinsky described the movement as 'an upside-down pyramid of fugues'. The opening section is the simplest, with entries (at Figs. 1, 2, and 3) occurring at the conventional pitch intervals (an alternation of tonic and dominant); a short episode (Figs. 4–5) leads to the second fugue, a vocal one, which is underpinned in the bass by a restatement of the orchestral fugue's main subject. There is a clear case of *stretto* at Fig. 10 (where the entries of the fugue subject 'pile up' on one another) and, just before the end of the extract, a solo trombone gives a further rhythmic transformation of the initial subject.

There are no grave problems in following this score, but the reader should be alert to the considerable number of precise performing instructions Stravinsky gives the executants. In the light of his statement quoted in the Introduction, this will be of no surprise. Note in particular the 'pause' marks for the first four notes of the initial subject, which serve to isolate the melodic line from any lyrical connotations and emphasize it as a progression of intervals. The detailed phrasing throughout is also of great importance, particularly as in many cases there is a complex 'counterpoint' of phrasing which adds to the textural variety.

©Copyright 1931 by Edition Russe de Musique
Copyright assigned 1947 to Boosey & Hawkes, Inc. for all countries
Revised Version © Copyright 1948 by Boosey & Hawkes, Inc.
Reprinted by permission of Boosey & Hawkes Music Publishers Ltd., London

9. E. Varèse: *Ionisation* (extract).

Edgard Varèse (1883–1965) wrote a number of compositions in the 1920s and 1930s which, in their use of rhythm and, consequently, percussion instruments, and their seeming disregard for precise pitch organization, were in many ways as revolutionary as the textural innovations of Debussy or the pitch organization of Schoenberg. None the less, he was not a complete 'original': influences may be discerned as diverse as the French pre-Classical tradition on the one hand (stretching back to the *organum* of Pérotin and the Notre Dame School), and on the other his contemporaries Debussy, Schoenberg, and Stravinsky.

He was, however, suspicious both of system and of immediate tradition. He never espoused twelve-note technique and his music is almost impenetrable to any but the most sophisticated analytical methods.

There can be no denying its power, however, and *Ionisation* (1934) is an excellent introduction. Scored for percussion orchestra, including piano, the work depends on rhythm and texture for its effect. (It is significant that the piano – the only traditional melodic/harmonic instrument in the ensemble – is used only sparingly.) The instruments used might not be considered unusual nowadays (although the sirens may still raise a laugh in some performances), but in 1934 the concept of an orchestra comprising only percussion was highly unusual. The distribution of the instruments used is shown at the end of this section.

The work opens quietly with predominantly sustained sounds (from the gong and tam-tams, sirens, cymbals, etc.), rather as if the piece is to grow from elemental beginnings. In this it may be compared to the opening of *La Mer* and *Jeux* by Debussy. The counterpart to a 'theme' begins in the bar before Fig. 1, where the side drum (*tambour militaire*) asserts its presence with a martial figure (accompanied by bongos) which is to be prominent in the opening sections of the work. Whether it is permissible to speak of the rhythmic figure at Fig. 1 as a 'theme substitute' is debatable: perhaps it is better simply to regard the side drum as the 'first violin' of the ensemble and, at least initially, to concentrate on that line. Listening to the whole work will show that pitched elements (the piano, for example) are merely inserted as additional sonorities: the real point of this piece remains rhythmic and timbral throughout.

The rhythmic complexity is largely a matter of durations and groupings rather than of tempo (which is constant throughout) or metre: the two changes of time signature may be heard as built-in rubato at the ends of phrases.

INSTRUMENTS

1. **Grande Cymbale Chinoise** large chinese cymbal
   Grosse Caisse (très grave) bass drum (very deep)
2. **Gong** gong
   Tam-tam clair high tam-tam
   Tam-tam grave low tam-tam
3. **2 Bongos (clair/grave)** high and low bongos
   Caisse Roulante tenor drum
   2 Grosses Caisses (moyenne/grave) middle and low bass drums
4. **Tambour militaire** military drum
   Caisse roulante tenor drum
5. **Sirène claire** high siren
   Tambour à corde snare drum
6. **Sirène grave** low siren
   Fouet Whip (also played by **13**.)
   Güiro scraper
7. **3 Blocs Chinois (clair/moyen/grave)** high, medium, and low chinese blocks
   Claves claves
   Triangle triangle
8. **Caisse claire (détimbrée)** high drum (without snare)
   2 Maracas (claire/grave) high and low maracas
9. **Tarole** tarole (small snare drum)
   Caisse claire high drum
   Cymbale suspendue suspended cymbal
10. **Grelots** small round bells
    Cymbales cymbals
11. **Güiro** scraper
    Castagnettes castanets
12. **Tambour de Basque** tambourine
    Enclumes anvil
13. **Piano** piano

TERMS

Baguettes Timbales (en peau) kettledrum sticks (on the skin)
Baguettes Timbales en feutre felt kettledrum sticks
Baguettes Timbales en bois wooden kettledrum sticks
Baguettes Tambour drumsticks **(pouce)** thumb

to Nicolas Slonimsky

IONISATION
(for Percussion Ensemble of 13 Players)

Edgard Varèse

COL. 8

© 1934 Colfranc Music Publishing Corporation, New York
By permission

50

10. A. Webern: String Quartet, Op. 28, First Movement

Anton Webern (1883–1945) is without doubt the composer who has most influenced avant-garde music since 1945. Composers such as Boulez and Stockhausen identify him, rather than Schoenberg, as their fundamental point of departure; Stravinsky, when he began to take up twelve-note composition in the 1950s, set the seal on Webern's reputation when he described him as 'a perpetual Pentecost for all who believe in music'.

The String Quartet, Op. 28 (1938), comes from Webern's most influential period, when he was producing the series of twelve-note instrumental works which formed the basis of his reputation among composers immediately after World War II. The note-row upon which the piece is based:

is both symmetrical (notes 9–12 are a transposition of notes 1–4; notes 5–8 an inverted transposition) and clearly evident in the opening of the work, albeit with octave displacement. Webern's musical language typically tends towards extreme economy and towards reduction of the possibilities available within twelve-note writing, as is seen here. Equally revealing is the fact that the letter names of the initial four notes spell out B–A–C–H when transposed up a minor third (in German, B=B♭; H=B♮); this gesture to the past reminds us that Webern regarded himself as part of a firm tradition – a logical continuation of the Austro-German music of the previous 200 years.

While it is beyond the scope of this short introduction to discuss in detail the structure of this highly complex movement, we might point out that most new sections are preceded by ritardandos (bars 14–15; 32; 47–8; 63–5; 77–8), and are often further emphasized by a change in tempo. Note the proliferation of expressive markings (virtually every note carries instructions for dynamics and method of attack) and the intensification of *Klangfarbenmelodie* techniques even compared with extract No. 4. Other features of the movement typical of Webern include the use of wide intervals, particularly the major seventh and minor ninth which had been characteristic of the expressionistic atonal works of his teacher Schoenberg, and the use of silence. The latter, in particular, serves to isolate and intensify individual sounds. Both these features can be seen clearly in the first bars of the extract.

As in Schoenberg's work, the eye and ear must learn to follow what is essentially a contrapuntal texture. As a start, the reader might follow through the various permutations of the basic B–A–C–H cell. The last vestiges of the homophonic melody and accompaniment still present in Mahler have now completely disappeared in favour of counterpoint. In this respect it is highly appropriate that Bach, the contrapuntal composer *par excellence*, stands behind the work.

INSTRUMENTS

Geige violin
Bratsche viola

TERMS

Mässig moderate
drängend pressing on
fliessender more flowing
wieder mässig moderate again
Dämpfer auf mute on
mit Dämpfer with mute
ohne Dämpfer without mute
wieder sehr mässig again very moderate
am Steg on the bridge
Dämpfer ab mute off
wieder nur mässig again only moderate
wieder fliessender u. noch drängend bis ⌒ again more flowing and still pressing on until ⌒

STREICHQUARTETT

ANTON WEBERN, Op. 28

Reprinted by permission of Universal Edition (London) Ltd.

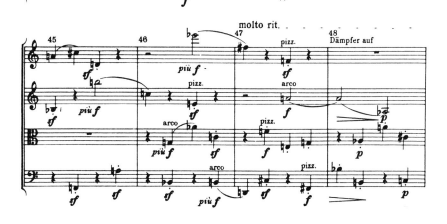

11. O. Messiaen: *Oiseaux exotiques*.

Olivier Messiaen (born 1908) has been a very influential figure in music since 1945. He has taught most of the leading figures of the avant-garde (Boulez and Stockhausen are two of the most prominent) and his book *Technique de mon langage musical** explains, in clear if didactic terms, the basis of his style.

His compositions have been dominated by a number of factors. First must stand his Roman Catholic faith which, as he writes in his books, prefaces to works, and programme notes, underpins everything he does. Then there is his interest in rhythm and the extension of European practice by the use of devices such as additive rhythms and 'rhythmic modes', often derived from Indian (Hindu) classical music. (A rhythmic mode is a pattern or grouping that recurs, either unchanging or developed additively, and unifies a composition.)

Related to both these preoccupations is an interest in bird-song. A keen amateur ornithologist, Messiaen often transcribes the sounds he has heard into his music, and in a whole series of works from *Le Merle noir* (1951) for flute and piano to the orchestral *Des Canyons aux étoiles* (1974), transcribed bird-song is either the major part of the material or at least forms a principal constituent.

Oiseaux exotiques (1956) is an interesting combination of these concerns. The preface includes some ornithological notes ('the hermit thrush, chest spotted with black, russet rail'), the names of the Hindu and Greek rhythms he has used, and 'a complete catalogue of the birds that sing in the score' — all forty-seven of them.

The bird-song is organized into the rhythmic patterns in an empirical way, but there is no doubt that it forms an essential part of the music — this is a composition, not an aviary. Whether it is an essential part of the score is another matter. Though Messiaen carefully indicates the various bird-songs, the best way to approach the piece is, as with any other conventionally notated work, to forget extra-musical associations and concentrate on the score as presented.

The work is scored for solo piano and a small orchestra of wind and percussion (Messiaen has probably omitted the strings for much the same reason as Stravinsky did in the *Symphony of Psalms*, feeling that their inherent expressiveness would obscure his intentions). The piano part is virtuosic in the extreme, with precise directions for all aspects of performance, especially phrasing (or method of attack).

The orchestral parts, also precisely notated, remain in $\frac{3}{8}$ throughout, primarily for purposes of co-ordination. The piano part, however, with no time signature in solo passages, follows the individual metrical implications of its phrases much more closely, with irregular bar-lengths. There is thus a distinction between the metrically free solo part and the ensemble, whose freedom is subordinate to the demands of co-ordination.

Some aspects of Messiaen's rhythmic technique are clear: on the 5th page, 4th 'system' (i.e. grouping of musical staves), several elements make up the call of the 'Grive des bois, d'Amérique'. The left-hand A♭-D♭ 'dyad' (chord of two notes) alternates between three and two occurrences. The six-note chord at the end of the second bar recurs at the end of the next and the next-but-one but the number of occurrences of this increases from three to six to eight. In this way Messiaen keeps some aspects (i.e. pitch material) of his music static while expanding the rhythm.

This example demonstrates in miniature one of his techniques of rhythmic development: another (on the 2nd page) is the alternation of static and mobile music: bar 4 (which repeats bar 1) is mobile — the glissandos giving an impression of movement — whereas bars 5 and 6 present repetitions of a chord (with durations of 3, 3, 3, and 2 semiquavers) which inevitably give the impression of 'no movement'.

Even more than Mahler, and just as much as Cage, Messiaen constructs in 'blocks'. There is no organic development in the traditional German sense: instead he has developed Debussy's techniques (cf. *Jeux*), presenting them in a stark uncompromising manner.

INSTRUMENTS
Petite Flûte piccolo
Petite Clarinette (Mi♭) E♭ clarinet
Caisse Claire snare drum

TERMS†
Presque lent almost slowly
Mainate hindou Common Mynah
sans timbre without snares
sec, dur dry, hard
sans péd. without pedal
vocifération implacable implacable cry of rage
Garrulaxe de l'Himalaya Himalayan Thrush
m.d./m.g. dessus right/left hand on top
timbre en bois wooden tone
dr./g. right/left
très sec very dry
hésitant hesitating
presser hurry on
Presque vif quite lively
Liothrix de Chine Red-billed Mesia
net, timbre en bois, proche du temple-block clear, wooden tone, close to that of the temple-block
Grive des bois, d'Amérique American Wood Thrush
Très modéré very moderately
un peu rubato, laissez longuement vibrer a little rubato, leave to resonate
éclatant, ensoleillé vividly lit up
Grive de Wilson Veery
Verdin de Malaisie Malayan Leafbird
Troupiale de Baltimore Baltimore Oriole
Grive de Californie Californian Thrasher

au chiffre 4 : les 2 éléments principaux sont: le Troupiale de Baltimore', par grand flûte, hautbois, clarinettes (la grande flûte doit s'entendre autant que le reste, elle est le soprano du groupe); la 'Grive de Californie', au xylophone. Ces 2 éléments sont également forts.
at figure 4 : the two principal elements are the 'Baltimore Oriole', on flute, oboe, and clarinets (the flute must be heard as much as the others, as it is the soprano of the group); the 'Californian Thrasher', on xylophone. These two elements are equally loud.

---

* (Paris, 1944); Eng. trans., *The Technique of my Musical Language*, by J. Satherfields (Paris, Leduc, 1957).
† Translations of bird names are taken from R. Sherlaw Johnson, *Messiaen* (London, Dent, 1975), pp. 199–208.

Oiseaux exotiques

(pour Piano solo et petit orchestre)

OLIVIER MESSIAEN

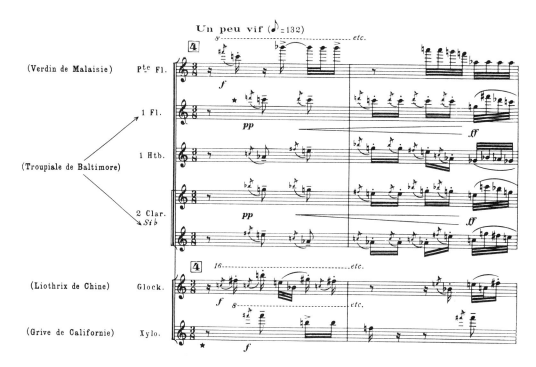

★ au chiffre [4] : les 2 éléments principaux sont: le "Troupiale de Baltimore", par grande flûte, hautbois, clarinettes (la grande flûte doit s'entendre autant que le reste, elle est le soprano du groupe); la "Grive de Californie", au xylophone. Ces 2 éléments sont également forts.

12. J. Cage: *Music for Marcel Duchamp* (extract).

'He's not a composer, he's an inventor — of genius' Schoenberg said of John Cage. Schoenberg, the staunch upholder of European tradition, taught Cage (born 1912) from 1934 and the various publications on or by Cage give some insight into this encounter.* Something of Cage's preoccupations and influences has been mentioned in the Introduction; here we see them in operation.

At first sight, the score of *Music for Marcel Duchamp* (1947) is unconventional as piano music. It is written in the alto clef (why? and especially, why for piano?) with the direction 'both pedals throughout', which gives a shimmering haze of sound broken only by the long rests Cage notates. Cage is thus 'typical' of the composers represented in these extracts in his use of texture and silence. There is not a great deal of two-part writing: indeed the extract might remind the reader of the spare textures of the piano music of Satie, one of the few earlier composers admired by Cage.

Given all these features, the look of the score — the sparse texture and the repetitions — is not so far from traditional notation. The sound, though, is made utterly different by the page of instructions which qualifies the conventional notation: Cage gives the performer precise details on how to 'prepare' the piano by inserting various materials between the strings.

This changes the sound, of course, but it does more than that. The relationship between the score-reader and the performance is altered: hitherto the instruction to the performer would also be an accurate anticipation for the reader; now the score retains instruction and even explanation (see Cage's note on the structure at the top of the page of instructions), but a listener can scarcely be prepared for the modification of expectation until a greater familiarity with the work and its sounds is achieved.

So here is a comparatively new kind of music — spare, stark in its juxtaposition of unrelated events (e.g. the change from the 3rd to the 4th page) and novel sound characteristics, and articulated by a notation in which verbal features are an essential complement to conventional note forms. The structure, consisting of 11 sections of 11 bars each, is also novel, even within the terms of twentieth-century music. Within these sections musical events occur in phrases (bar-lengths) according to Cage's note, so that the first 11 bars contain phrases of bar-lengths as follows: 2; 1 (repeated); 3; 1; 2 (silent bars); 1.

Compare: *Sonatas and Interludes* for prepared piano (London, Peters, composed 1946–8). See also J. Cage, *Notations* (New York, 1969).

* Especially *John Cage*, ed. R. Kostelanetz (New York, London, Allen Lane, 1971).

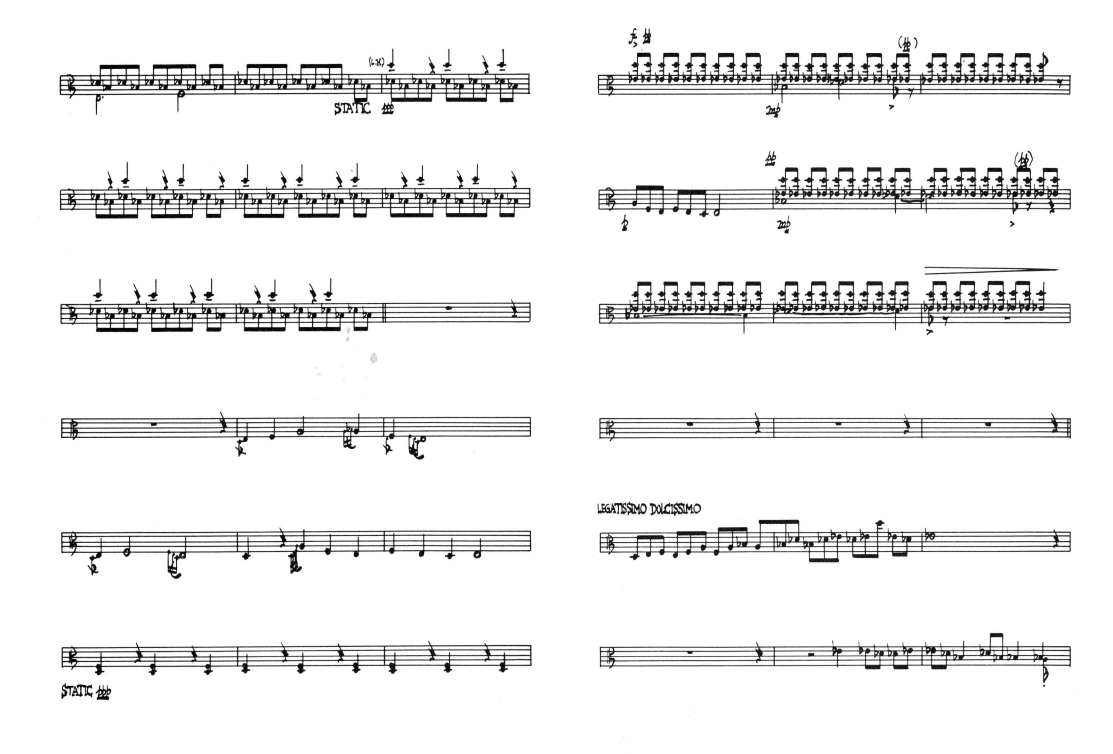

13. P. Boulez: *Le Marteau sans maître*, commentaire I de 'bourreaux de solitude'.

Pierre Boulez (born 1925) was one of the leaders of the generation of avant-garde composers who came to prominence after 1945. Taking Webern rather than Schoenberg as their point of departure, these composers (Stockhausen was another) developed 'total serialism' in which Schoenberg's twelve-note method was extended, by analogy, to duration, dynamics, and often other secondary areas.

Le Marteau sans maître (1955) is a cantata to words by the French surrealist poet René Char who used words for their quality of sound rather than their meaning in his poems. In nine movements, it sets three of his poems and surrounds them with instrumental music. The work is scored for contralto voice, alto flute, xylorimba (a xylophone with extended compass in the bass register to accommodate the low notes of the marimba), vibraphone, guitar, viola, and unpitched percussion.

The settings and associated instrumental movements are characterized by particular instrumental forces: 'bourreaux de solitude' and its 'commentaries' are marked by the use of xylorimba and percussion. This movement is the clearest in formal structure of the nine: our extract forms the 'A' section of a ternary form, recurring in telescoped form at the close of the movement. The 'B' section is completely different, with the use of hard sticks on the xylorimba, double-stopped viola playing pizzicato, and bongos producing a fragmented and irregular feeling in complete contrast to the measured pulse of the 'A' section.

The rigorous control exercised by Boulez over his performers may clearly be seen, although it is open to question whether the labelling of every single sound event with a dynamic marking does not veer towards affectation. The soft sticks of the xylorimba complement the *legato* of the alto flute (written a fourth lower than sounding pitch), which Boulez stresses should be at the limit of the capabilities of the player and the instrument.

Some of Boulez's notational innovations are also presented here. For example, the individual note values (e.g. in the alto flute part of bar 1) are retained to emphasize the individual nature of the notes even when they are beamed together (to indicate phrasing). This leads to obvious difficulties in performance — the complexities of the rhythmic language with its changes of metre are formidable — but Boulez aids the players to a certain extent by his subdivisions of the bar.

The texture is held together by the side drum (*tambour*, played at the rim), with its apparent regular pulsation; the alto flute seems to take the part of a voice exposing lines of a verse with short instrumental breaks in between.

The language of this extract is difficult: despite Boulez's directions for *legato*, fragmentation is still the basic feature of his music.* Score-readers approaching this style for the first time should perhaps concentrate on the flute and side-drum lines, later considering the relationship between the xylorimba and the viola (*alto*). The texture will then have the flute as soloist, with accompaniment from xylorimba and guitar and with a steady background from the side drum. We still have a traditional piece, traditionally notated — even if that tradition is now stretched into complexities which are daunting for the inexperienced.

INSTRUMENTS

Flûte en sol alto flute
Xylorimba, baguettes douces xylorimba, soft-headed sticks
Tambour sur cadre drum (on a stand)
Alto avec sourdine viola with mute
L'Alto pose l'archet pour jouer cette pièce the viola should put down the bow for this piece

TERMS

Lent. Les nuances seront exécutées 'ponctuellement' Slow. The nuances should be meticulously executed
Tempo rigoureusement exact jusqu'à l'indication contraire Tempo rigorously exact until there is an indication to the contrary
donner aux valeurs toute leur continuité, à la limite du legato give durations their full length, to the limit of the *legato*
Toutes les sonorités très équilibrées entre elles all sonorities very even
baguettes de caisse claire trés légères very light snare drum sticks

* See *Boulez* P. Griffiths, (London, OUP, 1978).

II.
commentaire I de «bourreaux de solitude»

14. K. Stockhausen: *Kontakte* (extract).

> What interests me most in 'electronic music' so far is the notation, the 'score'.
> (Stravinsky, 1958.)

In many ways Stockhausen (born 1928) has been the most sucessful composer of 'pure' electronic music (i.e. works in which a pre-recorded tape constitutes the entire piece). His *Gesang der Jünglinge* (1955) was the first work of this type to make a wide impression, and he followed it with several other similar pieces.

Simultaneously, however, his increasing interest in the act of performance led him to make versions of his 'pure' electronic works in which instrumentalists performed 'live'. *Kontakte* (1958–60) is the earliest example of this.

The composition is based on the idea of 'contacts' between electronic and instrumental sounds, and on 'contacts' within the tape part itself. It is organized in what Stockhausen has called 'moment-form', and this extract presents the first 'moment'. Stockhausen defines moment-form in this work as being either 'a state or a process, individual and self-regulated and able to sustain an independent existence'. In other words, this extract (2 minutes 10 seconds, as may be seen from the figures at the top of the score) is, at least in theory, able to stand on its own as a self-contained piece. Within it are six distinct events, labelled IA – IF; each is characterized by various different features from the dense electronic sound and activity in IB to the staccato noises of ID.

The tape part, which in live performance has to be balanced by a 'sound projectionist', is notated at the top of the score and is represented graphically – the composer's 'pictures' of his sounds.

In the full score 63 pages of explanation of the sounds and treatments precede the music.* They are a record of procedures, however, rather than instructions to the performer or aids to the listener, and should not be counted as part of the score.

The percussion is notated next. The percussionist has three types of sound at his disposal: metal (e.g. gongs), skin (e.g. snare-drum), and wood (e.g. templeblocks). At the bottom of the page is the piano part, the nearest to conventional notation. As in much contemporary music the semibreve-type means that the sound lasts for as long as the instruments can sustain it while quaver-like figures with a dash through them (e.g. at 24.3″ in the second 'block' of section IB) are to be played as fast as possible.

Dynamics and pedalling are notated conventionally (even the dynamics in the tape part) while the abandoning of 'bars' and time signatures is, of course, inevitable where the timing of all events is conditioned by a pre-determined tape. Of all the scores in the book this is the one that most demands reading while a recording is being played.

For a further discussion of the issues raised by the notation of electronic music, see H. Cole, *Sounds and Signs: aspects of musical notation* (London, OUP, 1974).

IA **am Rand entlang 1 x im Kreis streichen** stroke once around the rim
dünne Trommelstöcke thin drumstick
Rand rim
Klavier piano

IB **In allen Verschiedenes** each plays something different
Fuß foot
Vibraschlegel vibrastick

IC **Alternierend** alternating

ID **In allen das Gleiche** each plays the same
harter Gummi hard rubber
dick thick (drumstick)
Reproduktion nach der Handschrift des Komponisten a reproduction of the composer's manuscript

IE **Rotation I–II (etc.)** each player in turn I–II (etc.)
weich, aber *f* (Gummi) soft, but *forte* (rubber)
harte vibraschlegel hard vibrasticks

IF **Flutklang** (lit.) floodsound
hart (Kautschuk) india rubber

Kuppe rounded tip
Xyloschlegel xylostick
auf Maximum des elektr. Klanges the electronic sound as loud as possible
schnell mit Ellbogen fast, with elbow
bleibt xyloschlegel still with the xylostick
dicker Klöppel big clapper
äußerste Vorderkante on the outermost front rim

II **1 x langsam, etwas beschleunigend, im Kreis streichen** once slowly, speeding up a little, stroking in a circle
dünner Metallstab am Rand entlang, Abstand zwischen Hand und Berührungsstelle des Stabes kontinuierlich verkürzen (gliss. ↗) with thin metal rod following round the rim, continuously shorten the distance between the hand and the point the rod touches
im Kreis rühren hit in a circle

* See R. Maconie, *The works of Karlheinz Stockhausen* (London, Boyars, 1976).

Nr. 12 Kontakte

Karlheinz Stockhausen

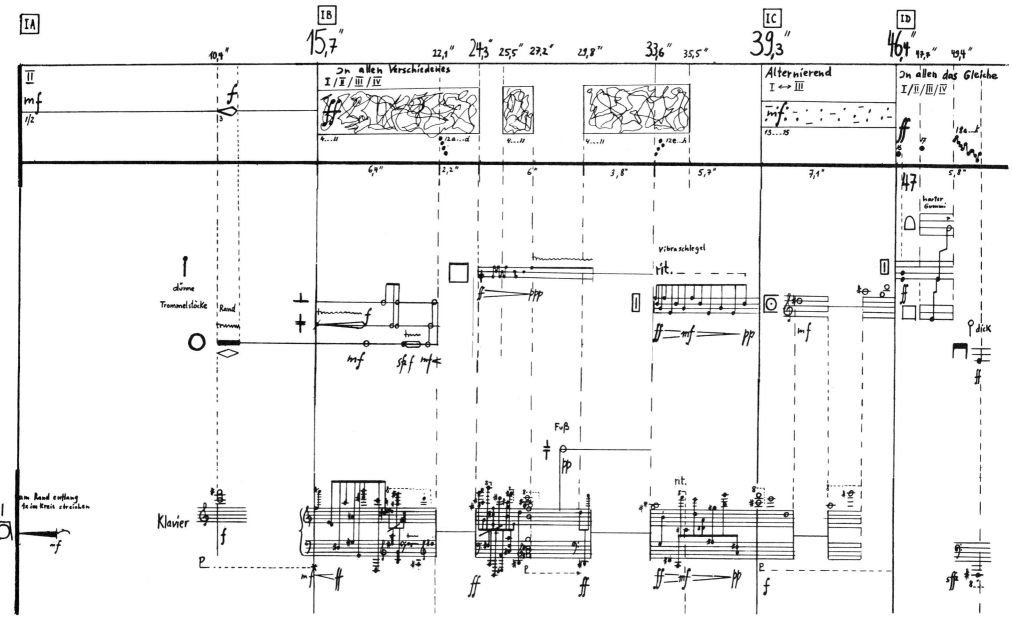

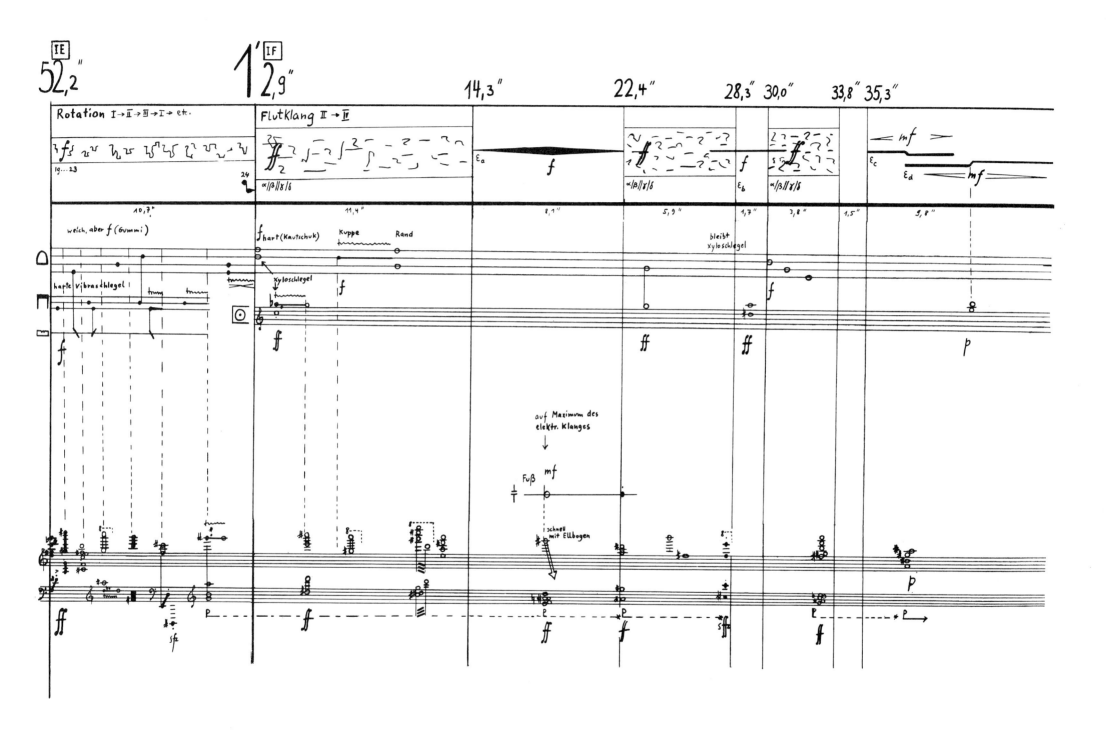

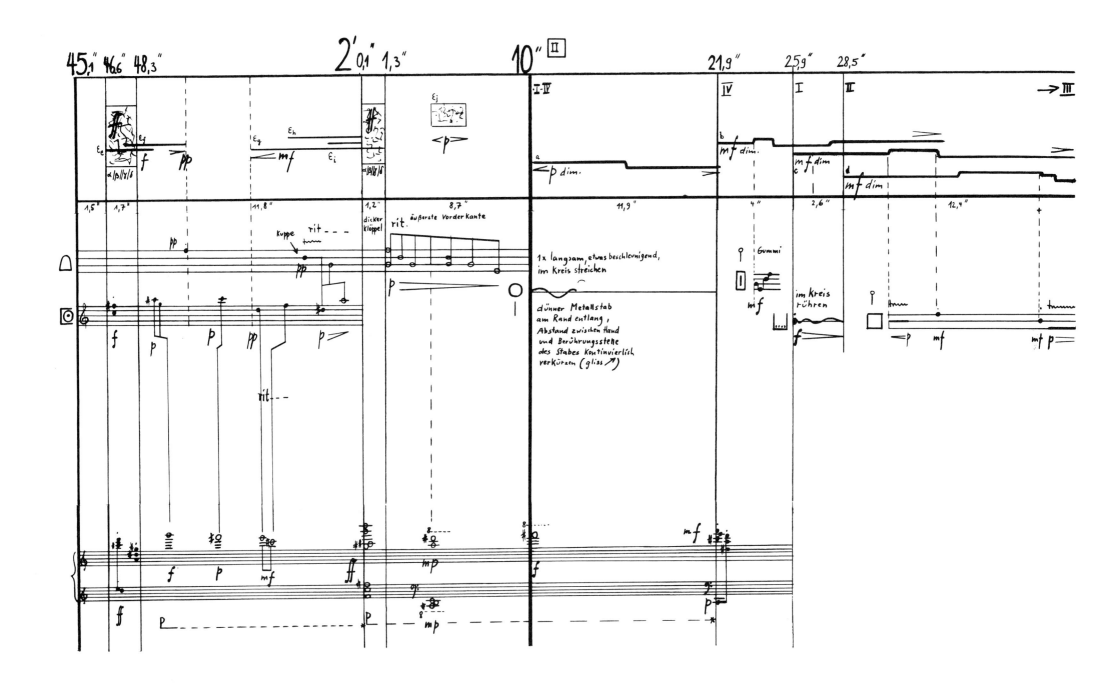

15. K. Penderecki: *Threnody: to the Victims of Hiroshima* (extract).

Krzysztof Penderecki (born 1933) first came to prominence in 1959 when he entered three works, anonymously, in a Polish national competition and won the first three prizes. After this he quickly gained an international reputation with pieces such as *Threnody* (1961), the *St Luke Passion* (1965), and the opera *The Devils of Loudun* (1968).

All these works mix stylistic elements, from simple melody and clear cadences to vast blocks of sound (e.g. mass string glissandos). Penderecki's music was once described as the human face of the avant-garde and his works are notable for their wide and direct appeal.

At first sight this score might seem as radical as Stockhausen's *Kontakte* or even Cardew's *Treatise*. But no new relationship between the score and the reader is set up: Penderecki simply extends the range of symbols used in order to prescribe the extended range of string sounds (sometimes of unspecific pitch).

Penderecki's music has been among the most radical in its extension of instrumental possibilities, particularly as regards strings. Perhaps taking his cue from Bartók (e.g. the 'night music' of the slow movement of *Music for Strings, Percussion and Celesta*), he has concerned himself more with texture than with pitch organization — with instrumental and orchestral colour rather than rhythm or harmony in their traditional senses. He may be compared with Lutosławski, Xenakis, and Ligeti in this, although each composer uses very different organizational methods.

Penderecki's are perhaps the most simple, immediate, and dramatically effective. This music is not intended to be descriptive (i.e. programmatic), and the title was added after the piece was composed.

The conductor's function is one of co-ordination rather than direction, since timings are strictly notated (see the lowest line of the score). There is little room for interpretation in this work, a fact which relates it, in a way, to the music of Stravinsky and Boulez (a comparison that would surprise all three composers).

The piece opens with staggered entries of all the strings playing the highest note possible. In the second and subsequent 'bars' Penderecki graphically represents different amounts of vibrato, while at Fig. 6 he requires the strings to produce a battery of effects (literally). All are described in the initial table.

Another graphic representation comes at Fig. 10, where Penderecki notates different types of cluster and glissando. His use of quartertones to produce an effect akin to 'noise' in electronic music may clearly be seen.

The notation of this work is thus an extension of conventional practice rather than a radical new development — except, perhaps, in the 'space-time' barring, where the length of each 'bar' is precisely stated (10 seconds, 15 seconds, etc.) rather than being determined by metre as in traditional rhythmic notation ($\frac{4}{4}$ = 4 crotchet beats per bar). This development of rhythmic notation is now a commonplace, with composers such as Berio and David Bedford, and composers of electronic music, making frequent use of it.

SKRÓTY i SYMBOLE	ABBREVIATIONS AND SYMBOLS		SIGNES D'ABREVIATION ET SYMBOLES	ABKÜRZUNGEN UND SYMBOLE
ordinario		ord.		
sul ponticello		s. p.		
sul tasto		s. t.		
col legno		c. l.		
legno battuto		l. batt.		
podwyższenie o ¼ tonu	raised by ¼ tone		hausse la note d'un quart de ton	Erhöhung um ¼ Ton
podwyższenie o ¾ tonu	raised by ¾ tone		hausse la note de trois quarts de ton	Erhöhung um ¾ Ton
obniżenie o ¼ tonu	lowered by ¼ tone		abaisse la note d'un quart de ton	Erniedrigung um ¼ Ton
obniżenie o ¾ tonu	lowered by ¾ tone		abaisse la note de trois quarts de ton	Erniedrigung um ¾ Ton
najwyższy dźwięk instrumentu (wysokość nieokreślona)	highest note of the instrument (indefinite pitch)		le son le plus aigu de l'instrument (hauteur indéterminée)	höchster Ton des Instrumentes (unbestimmte Tonhöhe)
grać między podstawkiem i strunnikiem	play between bridge and tailpiece		jouer entre le chevalet et le cordier	zwischen Steg und Saitenhalter spielen
arpeggio na 4 strunach za podstawkiem	arpeggio on 4 strings behind the bridge		arpège sur 4 cordes entre le chevalet et le cordier	Arpeggio zwischen Steg und Saitenhalter (4 Saiten)
grać na strunniku (arco)	play on tailpiece (arco)		jouer sur le cordier (arco)	auf dem Saitenhalter spielen (arco)
grać na podstawku	play on bridge		jouer sur le chevalet	auf dem Steg spielen
efekt perkusyjny: uderzać w górną płytę skrzypiec żabką lub czubkami palców	percussion effect: strike the upper sounding board of the violin with the nut or the finger-tips		effet de percussion: frapper la table de dessus du violon avec le talon de l'archet ou avec les bouts des doigt	Schlagzeugeffekt: mit dem Frosch oder mit Fingerspitze die Decke schlagen
kilka nieregularnych zmian smyczka	several irregular changes of bow		plusieurs changements d'archet irréguliers	mehrere unregelmäßige Bogenwechsel
molto vibrato	molto vibrato		molto vibrato	molto vibrato
bardzo wolne vibrato w obrębie ćwierćtonu, uzyskane przez przesuwanie palca	very slow vibrato with a ¼ tone frequency difference produced by sliding the finger		vibrato très lent à intervalle d'un quart de ton par le déplacement du doigt	sehr langsames Vibrato mit ¼-Ton-Frequenzdifferenz durch Fingerverschiebung
bardzo szybkie i nierytmizowane tremolo	very rapid not rhythmicized tremolo		trémolo très rapide, mais sans rythme précis	sehr schnelles, nicht rhytmisiertes Tremolo

Ofiarom Hiroszimy
TREN

KRZYSZTOF PENDERECKI
1959-1961

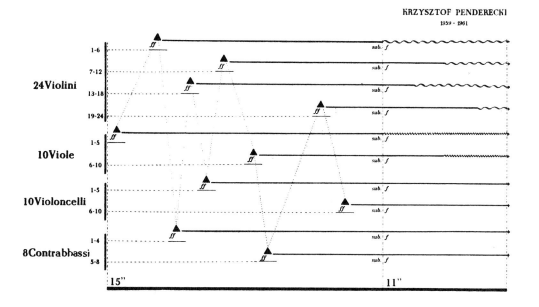
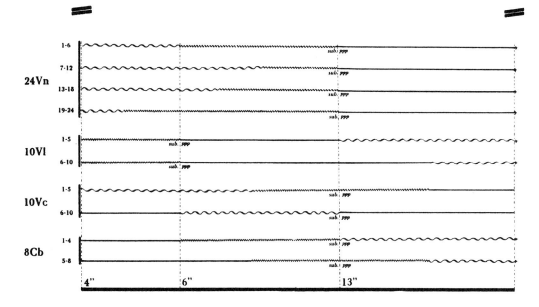
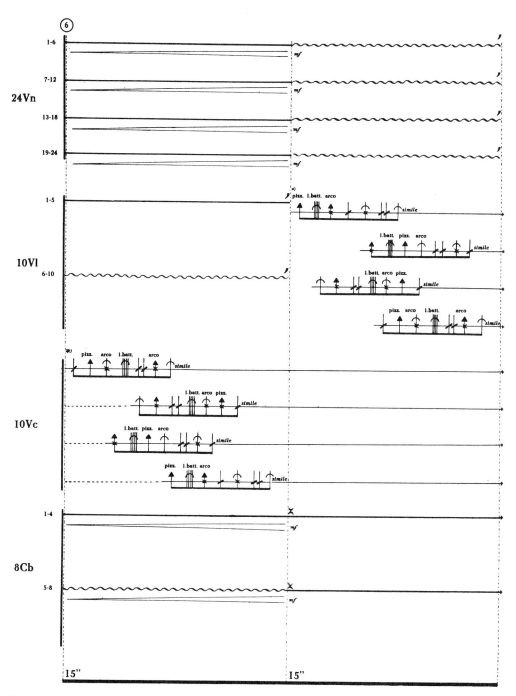

© 1961 Deshon Music Inc./PWM Edition for the World
Reproduced by kind permission of Belwin-Mills Music Limited, 250 Purley Way, Croydon CR9 4QD

*) Każdy instrumentalista wybiera sobie jedno z 4 podanych ugrupowań i realizuje je (w wyznaczonym odcinku czasu) tak szybko, jak tylko można.
Jeder Instrumentalist wählt eine der angegebenen 4 Gruppierungen und spielt sie (im bestimmten Zeitabschnitt) so schnell wie möglich.
Each instrumentalist chooses one of the 4 given groups and executes it (within a fixed space of time) as rapidly as possible.
Chaque exécutant choisit un des 4 groupements donnés et l'exécute (dans le segment de temps indiqué) aussi vite que possible.

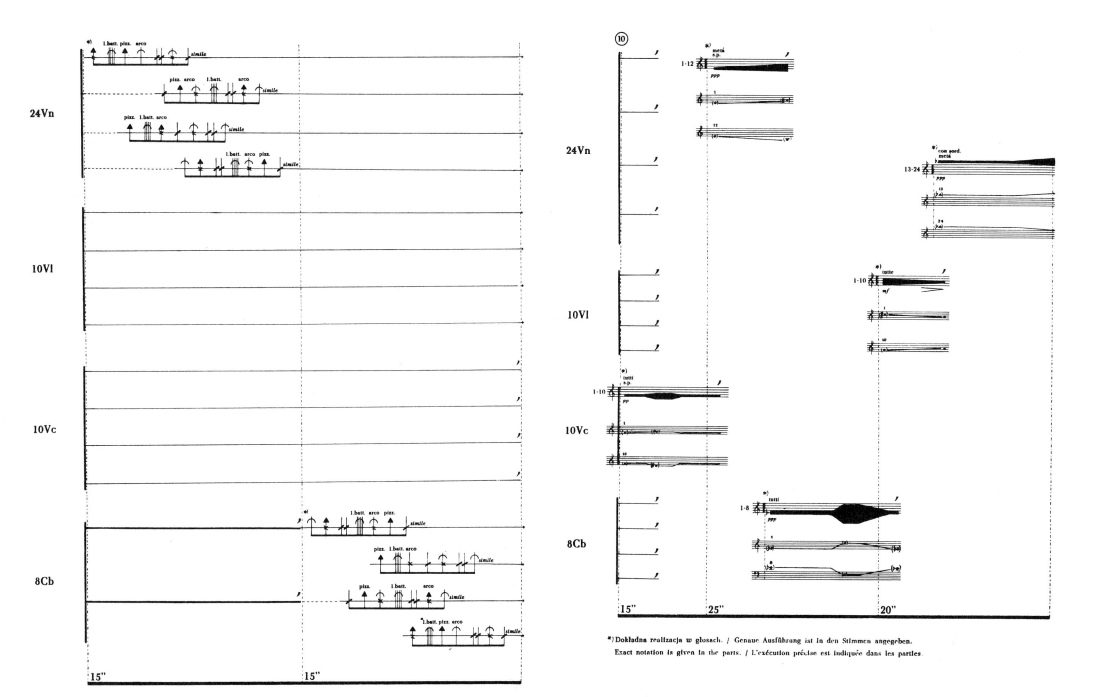

*) Dokładna realizacja w głosach. / Genaue Ausführung ist in den Stimmen angegeben.
Exact notation is given in the parts. / L'exécution précise est indiquée dans les parties.

*) patrz uwaga na s.6 / vgl. Anmerkung auf Seite 6 / cf. note on page 6 / voir note à la page 6

16. T. Riley: *In C.*

Terry Riley's *In C* (1964, subsequently revised) was one of the first 'systematic' pieces to capture the attention of the musical public. Taking the idea of adherence to a pre-determined scheme of composition from Cage's music, Riley's works proved considerably more accessible (because of the tonal musical material used) and, to a certain extent, less completely determined. Note, for example, in the performing instructions that each performer decides on the number of repetitions of each figure, the looseness of tempo direction (depending on the ensemble's ability), the layout of the ensemble, informality as to amplification, and so on.

Equally noticeable, however, are the constraints on this apparent freedom. All performers have to play the same figure, in strict tempo, and it is strongly recommended that the ensemble move, if not together, at least 'within sight' of one another. This preserves the harmonic feeling of the piece which, without perfect cadences, appears to move from C as follows:

C → G^7 → E minor → G^7 → G minor
I → V^7 → III → V^7 → V minor

The whole effect is kaleidoscopic, with gradually shifting areas of tonality perceived only in blurred outline.

The score-reader confronted with just the fragments in conventional notation will be able to follow the piece, though in an incomplete manner; they are easy enough to hear and to relate to the score. The 'explanation' contained in the performing instructions makes the work comprehensible (e.g. by stipulating the pulse) – even if Riley's language now seems rather dated! Riley also directs the procedure for moving from fragment to fragment and suggests the manner in which this should be done. One of the interesting things about the piece is the contradiction between the fragmentary appearance of the conventionally notated segments and the finished work – a continuous texture. This contradiction is, of course, the result of the verbal instructions.

In this piece, then, the written word is an essential adjunct to the notation: it even becomes a vital part of that notation. It does not qualify traditional graphics – it complements them as an essential aid to performer and listener alike. But, given this dual function, Riley's piece stands on the conventional side of the procedures outlined in the Appendix.

Performing Instructions

All performers play from the same part. There are 53 repeating figures, played in sequence. They are to be taken consecutively with each performer determining the number of times he repeats each figure before going on to the next. The pulse is traditionally played by a beautiful girl on the top two octaves (C's) of a grand piano. She must play loudly and keep strict tempo for the entire ensemble to follow. The tempo should be determined by how fast the ensemble can execute the smallest units (16th notes). All performers must strictly adhere to the tempo of the pulse. After the pulse has begun to sound each performer determines for himself when to enter on the first figure.

As a general rule the performers should remain within a compass of 4 or 5 figures of each other, occasionally trying to merge together in a unison. This means that although each performer is essentially free to repeat a figure as many times as he wishes, he must ultimately abide by the pace taken on by the majority of the ensemble. The ensemble should sit as close together as is comfortably possible, all performers radiating outward from the pulse who should be in the center. It is O.K. to amplify instruments that can't naturally play so loud, such as strings, flutes, harpsichords, etc. All parts should be played at the written pitch. It is generally O.K. to transpose up an octave. Transposing down an octave should be discouraged unless several performers are doing so and even then they should be extremely careful in choosing alignments and try to absorb the tendency to stick out.

Since performances often go over an hour each figure can easily be repeated for a minute or longer (performances could last days, months, a year – a figure for each week, with the closing one to start the new year). Don't be in a hurry to move from figure to figure. Stay on your part and keep repeating it, listening for how it is relating to what the rest of the ensemble is playing. If it sounds like everyone is playing in the same alignment of a figure, you may shift yours to create an opposing alignment.

Say that most of the ensemble is playing figure 12 like this:

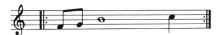

you may choose to align like this:

or this:

and so on.

In this way you have not 53 figures but literally thousands and it is part of the creative task of the ensemble to explore the different combinations.

Play at a good loud volume, but avoid letting your part get isolated from the others. Make all exits and entries as inconspicuous as possible. In order to play continuously without fatigue a figure can be repeated in groups of 4 or 5 with a rest included to form an overall pattern consisting of X number of repetitions plus a rest. The duration of the rest may be equal to the duration of the figure being repeated as illustrated below:

This makes the part more stable and symmetrical and allows it to groove better with the rest of the ensemble, especially if everyone else is stabilized in a pattern formation. However, any number of repetitions and any duration of rest is permitted. Changes should not be dramatic. The sequence should very gradually unfold. When each performer reaches figure 53 he should stay on that figure until the entire ensemble has arrived and a unison is achieved. The players can then gradually drop out, the pulse continuing a few moments after the rest of the ensemble has finished. If a performer comes to a figure he cannot play he may omit it. The ensemble should learn all figures in unison before attempting any combinations. It is essential that everyone play the figures correctly. Any number of instruments can play. In general the more players, the better it goes. Several keyboard instruments should be used as well as percussion instruments that are tuned such as marimbas, vibraphones and xylophones.

Appendix

Scores have traditionally been instructions to performers articulated through a symbolic language understood by both composers and performers. Since the late eighteenth century, scores have also been aids to listeners: those who could read music were able to follow the progress of a musical work.

The new methods of notation outlined in the previous three pieces do not call either of these functions into question, although the subsequent work of John Cage, for example, certainly does. However, with these two extracts the function of the score as information for both performers and listeners is abandoned, albeit in two contrasting ways. It remains to be seen how far these experiments point the way forward for our musical tradition.

17. C. Cardew: *Treatise* (*extract*).

Cornelius Cardew (1936–81) was the foremost composer in the British avant-garde during the 1960s and 1970s. After studying at the Royal Academy of Music, he worked with Stockhausen at the time the latter was composing *Carré*. After returning to England Cardew became interested in the ideas and notational devices of Cage, and *Treatise* may be seen as the synthesis of the two influences – the precise ordering of Stockhausen (there is a strict formal plan) as a background and the apparently free notation and interest in stimulating the performers' imagination derived from Cage. After *Treatise*, Cardew became dissatisfied with the isolation of avant-garde music and turned increasingly to expressing his socialist political commitment through music of a simpler kind.

Treatise is a score of 193 pages of graphic symbols. These symbols take musical notation as their point of departure but extend the purely visual aspects even further than some of Cage's later work. *Treatise* could be read simply as a succession of visual symbols deriving from musical notation: the only definitely musical aspect of the work is the pair of staves at the bottom of each page. No duration or instrumentation is specified.

The composer's plan was to write a score in which a certain number of graphic symbols would be varied, extended, and developed, much as themes were in Classical and Romantic music. The graphic ideas take on a life of their own, and there is thus a vast difference in the relationship between score and performance (and listening) here as compared with, for example, a nineteenth-century symphony where the score comprises instructions for performance and a record of the composer's *musical* intentions.

The score is of little use to the listener. There is no necessary correlation between what is seen and what is heard. There could be, however, if the performer(s) adopted a literal or evocative approach to the symbols; this shows the importance of the performer in the realization of *Treatise* and scores like it.

One of the main trends in music since 1945 has been to give the performer more control over the sounds they produce – hence the strange innovatory notations of Cage and Cardew. In *Treatise* it is up to the performer(s) to decide what the symbols mean and how they are going to interpret them. Two extreme possibilities are:

1. to use the score merely as pretext. Such a performance would be an improvisation and it would scarcely matter whether *Treatise* or anything else were the impetus for the performance;
2. to study the whole score and try to ascertain the composer's intentions as to his plan of the symbols and their development, to attach to each symbol a sound or sounds, and thus to follow the development of the symbols and vary the sound or sounds accordingly. For example, if it were decided that a dot would be a B♮ and that size of dot indicated volume, there would be quiet Bs on the left of the 1st page, slightly louder notes at the end of that page, and three considerably louder Bs on the next. Perhaps the flat in front of the block circle on the 2nd page would also alter the note, perhaps not.

One thing should be clear. This leaves a great deal to the performer in the making of decisions, but it also makes the concept of score-reading redundant other than in the most literal terms. For scores such as *Treatise* or even *Unbegrenzt* (the next extract) are visual documents only – to be viewed in the same way as works of visual art but with no background in the listener's aural experience to aid the reading. The conventions have gone.

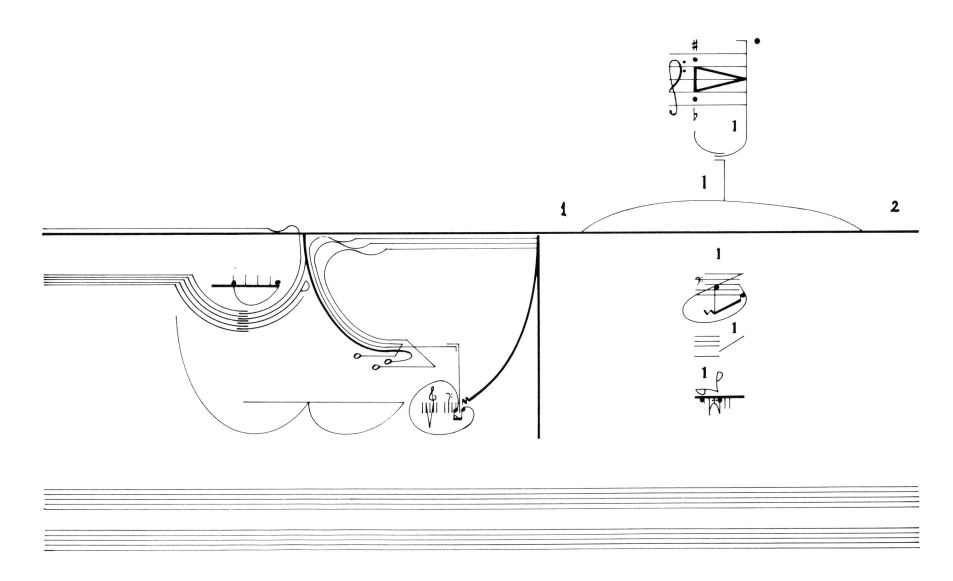

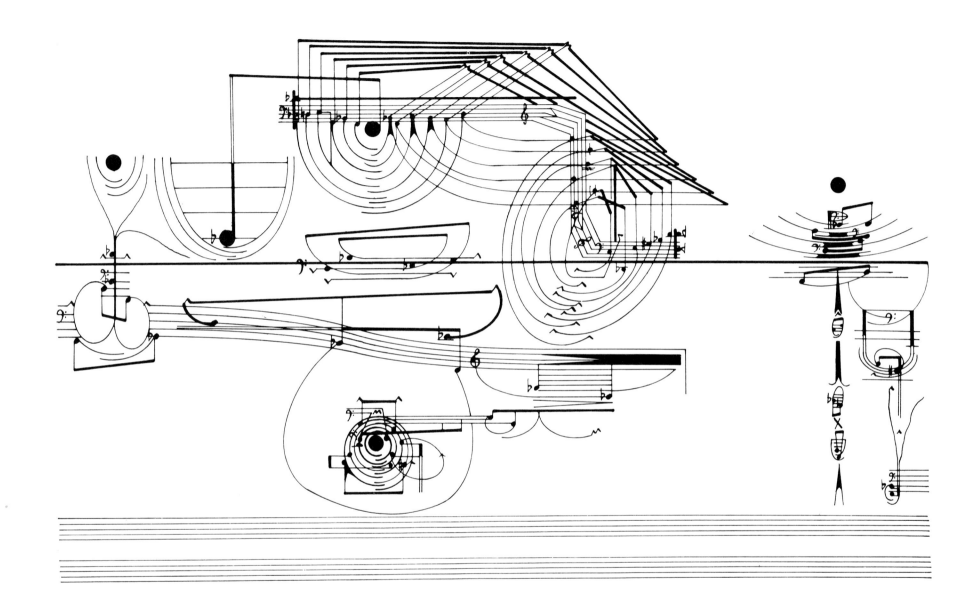

18. K. Stockhausen: *Aus den sieben Tagen*, 'Unbegrenzt' (*complete*).

Although Cardew's *Treatise* retains graphic symbols that are at least akin to conventional musical notation, Stockhausen dispenses with this entirely. His collection of texts 'From the Seven Days' is, however, clearly intended for musical realization, since:

1 the composer tells us so;
2 the composer (and others) perform the work as a piece of music (indeed, as a composition).

Stockhausen formed an ensemble to perform his less conventionally structured music (e.g. *Stop, Ylem*) and wrote fifteen texts for this ensemble in May 1968. The present extract is the only one of these texts to be accompanied by any graphic symbol.

At first glance, the reader may wonder how far something so utterly remote from conventional musical notation as the score of *Unbegrenzt* (Unlimited) deserves serious attention; but in fact the sketch does specify certain points. First Stockhausen directs that this is an ensemble piece, so that the interaction of the performers is essential. Secondly, he has ensured that the concentration of conscientious interpreters of the piece will be on the element of time – 'play a sound. . .infinite amount of time. . .'.

It may be objected that this is merely a stimulus to improvisation (where, by definition, a score is redundant), but Stockhausen carefully regulates all the performances in which he is involved (and, indeed, supervises recordings of them). In his activities as ensemble director, therefore, he is much less concerned than Cage with the liberation of the performer and we must accept that this is an attitude that underlies his music as well.

So at least one of the traditional functions of the score is preserved, in that the sketch is an introduction to the performer. It is very unspecific, however, and in that sense is remote from traditional musical conventions. If the listener is willing to suspend the normal score-reading (and rational) faculties, it may even serve as information. But it is clear that this is very different from the normal information presented in a score by Beethoven, Schoenberg, or even Boulez, and as such must be seen as a completely new departure, though different from Cardew's rather looser approach. That one must rely so radically on evidence external to the score for indications of Stockhausen's intentions demonstrates the remoteness of this notation from earlier practice.

für Ensemble

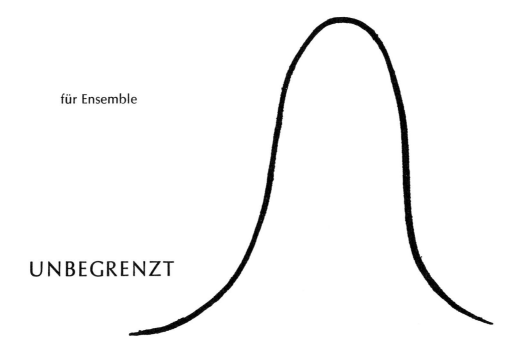

UNBEGRENZT

Spiele einen Ton
mit der Gewißheit
daß Du beliebig viel Zeit und Raum hast

© 1968 by Universal Edition A. G., Wien
English version © 1970 by Universal Edition A. G., Wien
Reproduced by permission of Universal Edition (London) Ltd.

8. Mai 1968

©

for ensemble

UNLIMITED

play a sound
with the certainty
that you have an infinite amount of time and space

may 8, 1968

©

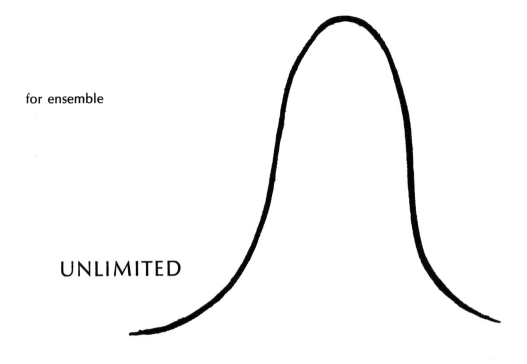